To Dennis Lindley and the memory of Clark Cockerham

Interpreting DNA Evidence

WITHDRAWN

LIVERPOOL
JOHN MOORES UNIVERSITY
AVRIL ROBARTS LRC
TEL. 0151 231 4022

LIVERPOOL JMU LIBRARY

3 1111 00930 4286

Although every care has been taken in the preparation of this book, the publisher and the authors make no representations, express or implied, with respect to the accuracy of the material it contains. In no event shall Sinauer Associates or the authors be liable for any indirect, incidental, or consequential damages arising from the use of this book.

INTERPRETING DNA EVIDENCE
Copyright © 1998 by Sinauer Associates, Inc.
All rights reserved. This book may not be reproduced in whole or in part without permission from the publisher. For information or to order, address:
Sinauer Associates, Inc., P.O. Box 407, Sunderland, MA, 01375-0407, U. S. A.

Internet: publish@sinauer.com
            www.sinauer.com

Library of Congress Cataloging-in-Publication Data
Evett, Ian
    Interpreting DNA evidence: statistical genetics for forensic scientists / Ian W. Evett and Bruce S. Weir.
        p.   cm.
    Includes bibliographic references and index.
    ISBN 0-87893-155-4 (paper)
    1. Forensic genetics.   2. Forensic genetics–Statistical methods
    3. Population genetics–Statistical methods
    I. Weir, B. S. (Bruce S.), 1943-   . II. Title
RA1057.5.E945 1998
614′.1—dc21   98-4081
                                                                      CIP

Printed in U. S. A.
5 4 3 2

LIVERPOOL
JOHN MOORES UNIVERSITY
AVRIL ROBARTS LRC
TEL. 0151 231 4022

Contents

Preface

In recent months, three different applications have brought home the widespread use and acceptance of DNA evidence. A jury in Santa Monica found a prominent football player responsible for the wrongful death of his former wife and her friend, a New York state crime laboratory was able to identify the remains of all 230 people who perished in an airplane crash, and scientists from the University of California at Davis were able to establish the parentage of the classic grape variety Cabernet Sauvignon. These very different uses of DNA rested on the ubiquity and uniformity of DNA throughout any living organism, and on the near identity of DNA between generations. Of almost equal importance in each case, however, was the numerical reasoning designed to show that matching DNA profiles provided strong evidence in favor of a single source of those profiles. It is this numerical reasoning with which we are concerned in this book. We have assumed some general knowledge of the field of DNA profiling, and would recommend the book *An Introduction to Forensic DNA Analysis* by Inman and Rudin for those without this knowledge.

Writing the book has been both a joy and a challenge. We have enjoyed laying out the foundations of a fascinating field, and we have been gratified by the response to early drafts from participants in the various short courses we have been teaching. We have struggled with the challenge of writing in a way that will be most useful for people who must confront DNA evidence in their professions but who do not have extensive training in statistical genetics. We have each been relatively recently introduced to the other's field, and we have enjoyed the challenge of accommodating each other's different backgrounds and philosophies.

We have benefited by good advice from many colleagues, and we wish to extend special thanks to Dennis Lindley, John Buckleton, Lindsey Foreman, Jim Lambert, Ian Painter, and Charles Brenner. We are very grateful for the comments we received from James Curran, Spencer Muse, Edward Buckler, Dahlia Nielsen, Christopher Basten, Stephanie Monks, Jennifer Shoemaker, and Dmitri Zaykin. Christopher Basten has helped with our LATEX problems. Andy Sinauer has remained a steady friend to us and our fields. We will welcome comments and suggestions, and we will display any corrections on the World Wide Web at http://www.stat.ncsu.edu (click on "Statistical Genetics"). We can be contacted via that address.

At several places in the text we have illustrated concepts by referring to data collected by the Forensic Science Service in the UK and by Cellmark Diagnostics in the US. We are grateful to both these organizations for their permission to use these data.

Gill Evett and Beth Weir have shown remarkable tolerance for our preoccupation with this project over the past few years. This brief mention of our gratitude is all too inadequate.

London and Raleigh
April, 1998

Foreword

Forensic science is experiencing a period of rapid change because of the dramatic evolution of DNA profiling. The sensitivity and discrimination of techniques now in routine use in many countries were undreamed of ten years ago. DNA has entered the vocabulary of the man in the street, perhaps not so much because of the beautiful work of those such as Watson and Crick as more because of the dramatic impact DNA profiling has had on crime detection.

One of the consequences of this new technology for the forensic scientist is that the strength of the evidence presented at court is usually expressed numerically. Whereas that has been the case for conventional serological techniques for decades, there are now two differences: the scale of implementation of the new methods and the enormous power of the evidence. A match between the profiles of a biological sample from the scene of a crime and that of a suspect has now been shown in many courts in various jurisdictions to have sufficient probative value to convince a jury that the sample came from the suspect, without the need for nonscientific corroborating evidence. The question often asked of a DNA profile is "Is it as good as a fingerprint?" We will meet this question later in the book (like many apparently simple questions, it does not have a simple answer!), but here it gives us an opportunity to reflect on a fascinating paradox.

Many will share the view that DNA profiling is the greatest advance in forensic science since the acceptance of fingerprint identifications by the courts at the turn of the century. Since that time, hundreds of thousands of opinions have been given by fingerprint experts. A fingerprint opinion is never a shade of grey–it is a categorical "those two marks were made by the same finger." This is accepted by courts throughout the world, in most cases without challenge, and the original introduction of this kind of evidence was, apparently, fairly painless. The statistical justification for fingerprint identification was rather sketchy and mainly theoretical.

DNA profiling, on the other hand, received something of a baptism by fire. For a few years, it was conventional to refer to the "DNA controversy," and in some countries there have been long and sometimes bitter courtroom confrontations. The controversy has, in turn, contributed to an explosion in the literature on the subject. Hundreds of papers have been published on DNA profiling statistics, many dealing with data collection, many others dealing with theoretical considerations of probability, statistics, and genetics. Although there were times when the controversy became acrimonious and testifying was unusually stressful, most would now agree that this extended debate has been good for the science. We know that DNA profiling is here to stay and that the statistics of the current techniques have been, in the main, established as robust. We now can say that we understand far more about the statistics of DNA profiles than about any other forensic technique–including fingerprints!

But another consequence is that there is now an appreciable body of knowledge with which the forensic practitioner must be comfortable if he or she is to report results and give evidence. This book is written with the aim of helping the forensic scientist who works in the DNA profiling field to gain sufficient knowledge of the statistical and genetic issues to report cases and to testify competently. We hope that there will also be much in the book that will interest other groups, such as lawyers and judges, as well as researchers in other fields.

Many forensic scientists engaged in DNA profiling have backgrounds that are strong in the biological sciences but relatively weak in mathematics. We consider that our "core reader" will be a forensic scientist with a degree in one of the biological sciences; thus he or she will be familiar with basic genetics, though we do review the necessary terms and concepts. In our experience, such scientists are often uncomfortable with statistics and so we have deliberately taken a gentle pace over the first three chapters. We have not assumed knowledge of calculus, but we do assume that our readers will be familiar with such mathematical ideas as logarithms, exponentiation, and summation. We must also recognize that there will be other readers with stronger mathematical backgrounds who are interested in a deeper coverage of some of the issues. For the most part, we have worked at keeping the mathematics as simple as possible, and our main aim has been to expose the underlying principles. However, from time to time, we have included more mathematical topics, and wherever possible we have enclosed these in boxes so that the reader who so wishes can skip them.

Because the field is changing rapidly, it has been rather difficult for us to decide what subjects to include and what to leave out. After considerable deliberation we decided to concentrate on the current generation of polymerase chain reaction (PCR) nuclear DNA-based profiling systems. It follows that we have included nothing on the subject of treating measurement error in comparing profiles, so readers will search in vain for a discussion of match/binning. We have not included anything on mitochondrial DNA statistics. That is rapidly becoming a subject in its own right and we have decided, with some reluctance, to defer any treatment of it to a future edition.

We devote all of Chapter 1 to an explanation of basic probability theory because we regard a good grounding in probability to be an essential prerequisite to an understanding of the problems of forensic inference. In Chapter 2 we show how the Bayesian approach to inference provides a logical and coherent framework for interpreting forensic transfer evidence. In particular, we show how the likelihood ratio is of central importance for forensic interpretation. Chapter 3 covers those topics in basic statistics that are necessary for our purposes; in particular, the theory underlying the estimation of allele proportions, and the basics of classical statistical testing of independence hypotheses. In this chapter we introduce a simple forensic example, based on a rape offense in the imaginary Gotham City, and we use this to discuss both the statistical and population genetics issues involved in assessing the strength of the evidence when the crime profile matches that of a suspect.

Chapter 4 introduces the concepts of population genetics and develops those ideas that are relevant to forensic science. In Chapter 5, we combine the statistical ideas of Chapter 3 with the population genetics from Chapter 4. Chapter 6 is a discussion of inference in cases of disputed parentage. We consider simple paternity and also more complicated situations such as incestuous paternity and identification of human remains. In Chapter 7 we consider the interpretation of cases of profiles of mixtures, illustrated by several examples. In Chapter 8 we return to the Gotham City example introduced in Chapter 3 to illustrate how match probabilities should be calculated, and in Chapter 9 we continue using the example to give our views on how the interpretation of matching profiles should be explained in a statement or formal report. Finally, we talk about the presentation of evidence in court, with particular reference to recent Appeal Court judgments in the UK.

We trust that the reader will appreciate our strategy of using boxes to separate the more mathematical passages. However, we must warn that this approach was inadequate in a few places: indeed, we should have liked to have much of the later sections of Chapters 4 and 5 in boxes!

Chapter 1

Probability Theory

INTRODUCTION

Events, Hypotheses, and Propositions

The word *event* is often used in the context of probability theory. Basically, we take it to mean any occurrence, past, present, or future, in which we have an interest. Sometimes, in real-life applications when dealing with nonscientists, it appears a little strange. Thus, for example, it is not customary for a lawyer to refer to the event that the defendant assaulted the victim. There are other words we can use in our attempts to understand the world: as scientists we might talk about the truth of a *hypothesis* or a *proposition*. In the legal context, it is common to use the word *allegation*. We will use each of these terms in what follows to mean essentially the same thing; on each occasion we will pick the word that best seems to suit the context.

Randomness

The word *random* is used frequently in basic statistics textbooks and also in the forensic context, particularly when the concept of "a person selected at random" is invoked. We will take some time to explain our understanding of the word and the meaning we assign to it when we use it in this book.

When we are talking about DNA types and we talk about "selecting a man at random" we mean that we are going to pick him in such a way as to be as uncertain as possible about his blood type. Another meaning of "random" in this context is selecting a man in such a way that all men have the same chance of being selected. This, however, is a more abstract way of looking at things, particularly when you imagine the practical problems of actually doing such a selection. Later, we will be talking about human populations in which there is random mating, and in that sense we mean that each person picks a mate in such

1

a way that all of the other members of the population have the same chance of being selected. Clearly this is an abstract idea.

Consider another example. Imagine that we have a chicken's egg before us and we are interested in its weight. Of course, we know something about its weight: it's certainly greater than a milligram and it's certainly less than a kilogram. But as we attempt to make increasingly precise statements, we become increasingly uncertain about their truth. If we can benefit from the experience of having weighed eggs on previous occasions, then we might be able to say that the weight of this particular egg is about 60 g. This represents our best guess, on the information available to us: it might be 50 g, it might be 70 g. The weight of the egg is a random quantity for us.

We can improve our knowledge by actually weighing the egg. Imagine that we use a kitchen scale and it reads 55 g. We now know more but we still can't be certain of the weight. For example, we are not sure about how well the scales have been calibrated, and in any case the scales are graduated only to the nearest 5 g. We now believe that the weight lies between 50 g and 60 g with 55 g our best guess: but the weight itself is still for us a random quantity. The reading on the scales represents data (strictly speaking, a single *datum*) that has enabled us to update our uncertainty about the weight of the egg. The science of statistics is all about using data in this way.

To illustrate our next point, let us consider a simple family board game in which the move of each player in turn is determined by his or her rolling of a single six-sided die. How is that rolling to be done? We could allow each player to hold the die in the hand and place it on the board, but we don't do that because we know that an unscrupulous player, such as five-year-old Suzy, may sneak a look at the die and place it in a way that favors the outcome she needs. So we use a cup, and each player is required to place the die in the cup and tip it out of the cup onto the table. But still the determined unscrupulous player can bend things in his or her favor by placing the die carefully in the cup and sliding it down the side onto the board: the required outcome is not guaranteed but this method appears to favor that which the player seeks, and the other players might feel that it was unfair. The next step then is for the players to agree on a means of manipulating the cup and of ejecting the die from it. What is their requirement? They want the outcome to be fair, and in this sense it means that the outcome should not favor any one player. How do they achieve this requirement? By making the shaking and tossing processes so chaotic as to make the outcome as unpredictable, or as uncertain, as possible. Note that when we were talking about the egg we were attempting to reduce our uncertainty about something (the weight of the egg), but in this second example we are trying to *maximize* our uncertainty about something (the score that the die will show). If we asked our players what they meant by *fair*, in the present context, they would say that all of the six possible outcomes should have the same probability. Note that we have slipped in this last word before we get to the section on probability and we are not yet talking mathematically. But we don't apologize because we don't think that mathematical concepts are needed here. The players might have used other terms such as "likelihood" or "chance" but their meaning would have been clear to each other and to us. In this context, fairness is equated to the same chance for each outcome.

The process of shaking and tossing is called *randomization* by statisticians; its sole purpose (other than perhaps to heighten the suspense of the players) is to maximize uncertainty, and this, it turns out, is the best way of achieving fairness. It can be shown using mathematics that maximum uncertainty in the present context is equivalent to the same chance for each outcome; we are not going to give this proof because it is a digression from the main theme of the book, and in our experience most people seem to find our assertion intuitively reasonable.

It is worth saying a little about popular misconceptions relating to randomization, probability, and chance (we are using the latter two words interchangeably at present because in everyday language they are synonymous). There is a tendency to think that the achievement of our belief in equiprobable outcomes in the die throwing example is some physical property of the die. Of course, it is in part because we first establish that the die has a different number on each of its six faces and that a piece of lead has not been craftily inserted immediately behind one face (there wouldn't be much point in using such a die in a board game but it could give an edge in a gambling game such as craps). Given those assurances, our beliefs about the outcome are a consequence of the state of uncertainty we have deliberately created. There is no mysterious physical mechanism–though there is a widespread tendency among both laymen and scientists to speak and act as though there is. Jaynes (1989) calls this the *mind projection fallacy*. It is a common idea to say that it is the mechanism of chance which ensures that in the long run we end up with equal numbers of each of the six outcomes. But there is no mechanism, no mysterious natural force field governing the behavior of the die. We end up with equal numbers because we have deliberately maximized our uncertainties. This is the essence of randomization.

We sometimes hear of "the laws of chance" or "the law of averages" as though these are natural laws that govern the behavior of the universe. They are not. They are simply the consequences of our efforts to maximize uncertainty. We will later be talking about the laws of probability, but there we mean something that has a completely different sense from the foregoing popular conception. We will be using probability as a mathematical tool for understanding how we may reason logically in the face of uncertainty. We arrive at the laws of probability in such a way as to make the tools work in such a sensible fashion: they are rather different in nature from, for example, Newton's laws of motion, which are intended to describe aspects of how the natural universe behaves.

We conclude this discussion by referring to an essay by noted mathematician Mark Kac (Kac 1983). Kac pointed out how difficult it is to define randomness and concluded:

> The discussion of randomness belongs to the foundations of statistical methodology and its applicability to empirical sciences. Fortunately, the upper reaches of science are as insensitive to such basic questions as they are to all sorts of other philosophical concerns. Therefore, whatever your views and beliefs on randomness–and they are more likely than not untenable–no great harm will come to you. If the discipline you practice is sufficiently robust, it contains enough checks and balances to keep you from committing errors that might come from the mistaken belief that you *really* know what "random" is.

We believe that the discipline covered in this book is indeed sufficiently robust that the genotypes of samples of people, chosen without knowledge of their genotypes by forensic scientists, can be regarded as random.

PROBABILITY

Uncertainty follows from a deficiency of information and our lives are characterized by decisions that must be made in the face of uncertainty. The issues involved in most real-life decisions are far too complex and fleeting to justify any attempt at logical and numerical analysis. However, there are areas where some progress can be made. Indeed, in the world of science it might be argued that there is a need, wherever possible, to ensure that one's reasoning is governed by logic and quantification rather than emotion and intuition. This book is concerned with one area of science where there is such a perception.

Probability theory has a history that extends over 200 years, taking its origin in the study of games of chance–the original motivation presumably being that a person who understands decision making in the face of uncertainty will gain an edge over someone who does not. But there have been several different ways of approaching the problem. Each one starts with a different view of probability and how it is defined. Remember that we are talking about a mathematical concept rather than a concrete facet of the natural universe. At the end of the day, the best definition of probability is going to be the one that will take us furthest toward the goal of rational thought, without our losing sight of the real-world problems we are attempting to solve.

Probability Based on Equiprobable Outcomes

Here is an early definition of probability. Think of a hypothetical experiment with several outcomes–the die rolling in the previous section is such an experiment. Now think of an event that is true if one of the outcomes happens. The event "the score is even" is true if the die shows a 2, 4, or 6; the event "the score is greater than 2" is true if the die shows a 3, 4, 5, or 6. Then, if all outcomes are equally probable, the probability of an event H is defined by

$$\text{Probability of } H \; = \; \frac{\text{Number of outcomes favorable to } H}{\text{Total number of outcomes}}$$

What do we make of this? First of all, it is a definition that can be of considerable use in analyzing many complicated problems, particularly those involving games of chance. It has some serious limitations, however. First, note that it is a definition of probability yet it contains the stipulation that the outcomes must be "equally probable," so the definition is circular. Second, it is restricted in its range of application; indeed, it is useless for most real-life situations. In a criminal trial, for example, a court is concerned with the uncertain event that the defendant committed the crime: there is no possibility of envisioning a number of equally probable outcomes in this situation. We face the same difficulty if we try to invoke the definition to answer a question of the kind "what is the probability that it will rain tomorrow?"

Long-Run Frequency

Defining probability as a long-run frequency is the basis of a school of statistical thought known as *frequentist* or *classical*. If we wish to talk about some event H, then it needs to be regarded as the outcome of an experiment that can, in principle at least, be carried out a large number of times. The outcome is assigned a numerical value, or *random variable*, which in this case can take two values, say 1 if H is true and 0 if H is false. We are interested in the number of times that the random variable takes a value equivalent to H being true. Let us carry out N identical experiments (e.g., roll a die N times). If we observe the event H (e.g., the score is an even number) occurring n times then we define the probability of H as the limit of n/N as N approaches infinity. So the probability of H can be determined only by experiment.

This definition can be very useful and, as we have said, most of the modern science of statistics has been built on this foundation. We will be using probabilities assigned on the basis of frequencies a lot later in the book. A useful distinction is to call probability assigned in this way "chance."

However, the frequency definition has limitations. It is intended to make statements about the behavior of random variables; indeed, frequentist probabilities can be applied only to random variables, and the concept of a very long run of random variables is central. This type of probability is then quite different in nature from the probabilities that we talk about in everyday life. If we ask "What is the probability of life on Mars?" then there is no useful purpose in attempting to visualize an infinite collection of planets indistinguishable from Mars. Indeed, this question cannot be answered with a frequentist probability. We face a similar problem when we talk about court cases. The question "What is the probability that this is the defendant's semen?" has only two answers in the frequentist sense: it either is or it isn't, so the probability is either one or zero.

Subjective Probability

A growing number of statisticians have been dissatisfied with the frequency definition of probability and its inferential consequences, and this brings us to the second main school of thought, called *Bayesian* or *subjectivist.* This school recognizes probability simply as a measure of our degree of belief. Although this might sound simple, there are a number of subtle arguments that need to be established before a system of mathematical rigor can be built upon it. We are not going to attempt to get involved in those arguments but refer the interested reader to fundamental works such as those by O'Hagan (1994) and Bernardo and Smith (1994).

A measurement system should have some kind of calibration standard, and this can be achieved in the following manner as described by Lindley (1991). Let us imagine that we are thinking about whether it will rain tomorrow afternoon at the time we are planning a barbecue. Denote by R the event that rain will spoil tomorrow's barbecue. Imagine a large opaque container that holds 100 balls. They are indistinguishable in size, weight, and shape but there are two colors: black and white. The container is shaken vigorously so that every ball has the same chance of being drawn. We are going to dip in and draw out a ball; we are interested in the probability of the event B that the ball is black. So we now have two uncertain events under consideration: R (rain) and B (black ball). Imagine now that we are going to be given a cash prize if either we correctly predict rain or if we correctly predict a black ball. So we have to choose: which gives us the better chance of winning a prize, predicting R or predicting B? To help us choose the wager we are told the number b of black balls in the container. Now if b is high, say over 90, then the better wager may appear to be that on B (unless we are thinking of a barbecue at a completely unsuitable time of the year). On the other hand, if b is less than 10, then the rain wager may appear preferable.

Let us try to think of that value of b at which we are completely indifferent to which wager we are going to choose: assume that we decide that this happens at about $b = 20$. Then our probability of R is $20/100 = 0.2$.

It is this concept of probability that will form the basis of the discussion for the rest of this chapter, although the formal discussion about the laws of probability also holds for frequentist probabilities. Note that as we are talking about subjective (or personal) probabilities we will not talk about them as though they had an independent determinate existence. Instead of talking about determining a probability as we would if we had a frequency definition, we will talk about *assigning* a probability. Given a set of circumstances, you will assign your probability to a particular event, but another person may assign a different probability. Is it a weakness that different people may assign different probabilities to the same event? Not at all–it is an inescapable feature of the real world. We discuss this issue in the next section when we talk about conditioning.

All Probabilities Are Conditional

If you respond "one-half" to the question "What is the probability that this coin will land showing a head?" then think again. Does the coin have two different sides? How is it to be dropped? Has it been loaded in any way? Only if the answers to these questions are appropriate does it make sense to assign one-half to the probability of a head. But the most important point here is that the probability that we assign depends on what we know: every probability is *conditional*.

In our family board game, if we ask "What is the probability the next throw will be a 3?" then our answer will be conditioned on the following: the die has six sides numbered 1 to 6; the die has not been loaded; the tossing is designed to maximize uncertainty. Only then can we agree on the answer $1/6$.

Recall the example of the egg. We weren't actually talking about probability at that time, but we were talking about uncertainty. Our first judgment about the weight of the egg was conditioned by our previous experience about eggs. We could, if we had been asked, have assigned a probability to a proposition of the kind "the weight exceeds 110 g." Later, we learned something new–an item of data–which changed our state of uncertainty and would have undoubtedly changed our probability for the proposition in the previous sentence. The *conditioning* has changed.

We will be saying much more later about how probabilities change in the light of new data, but at this point we need to digress briefly to talk about three words we will be using frequently: data, information, and evidence. Some writers would not distinguish between these three in a discussion of probability, arguing that probabilities can always be viewed as being conditioned by data.

But we believe that understanding can be eased by making distinctions in the following way. We will use *data* when we are referring to observations that can be quantified in some way: for example, they might be a set of genotypes from samples examined by a scientist. We will use *information* to refer to things that are not so easily quantifiable, such as an eyewitness report that the offender was Caucasian. This word has a more general meaning, and if we have a collection of information that includes some data, such as a report that there were three offenders, then we will use the more general term. The word *evidence* will be used in a still more general sense to include both data and information, particularly when we are talking about preparing a statement or presenting results in court. The distinctions between the three are not hard and fast and the reader should not feel confused if we sometimes use one of the three rather than another that appears more suitable in a given situation.

In our experience, when scientists disagree about the probability of a proposition, then it is often because they have different conditioning. It may be a simple matter of knowledge; for example, in speculating about the probability of rain tomorrow, one person may have better local knowledge of conditions than another. In scientific arguments, one proponent might have a slightly different model for reality than another does. It may be a matter of information: in the weather example, one may have heard a different weather forecast from the other. This is the reason that it is necessary always to be as clear as possible about the conditioning.

Notation

We have now reached the stage at which we need to introduce some notation. Choosing notation is always a compromise, but without it arguments become impossibly verbose and even harder to understand. Comprehension can be aided by a carefully chosen notation.

This is our first piece of notation: $\Pr(H|E)$, that is shorthand for "the probability of H given E." Here, H is some event or proposition about which we are uncertain, and E is the evidence (information and/or data) that we have in relation to H. If H is an event, then, depending on context, we might read $\Pr(H|E)$ as

- the probability that H has occurred,

- the probability that H will occur, or

- the probability that H is true.

We will also use $\bar{H}$ to denote the *complement* of H. Then $\Pr(\bar{H}|E)$ denotes

of Asian Indians are type 8,9.3; then it is reasonable to assign $\Pr(K|J) = 0.048$. So

$$
\begin{aligned}
\Pr(J \text{ and } K) &= \Pr(J)\Pr(K|J) \\
&= 0.25 \times 0.048 = 0.012
\end{aligned}
$$

We return to this example later.

Exercise 1.2   In another hypothetical country, 80% of the registered voters are Caucasian. Of the Caucasian voters, 20% inhabit the highlands and the remainder the lowlands. Among the Caucasian highlanders, 75% speak an ancient Celtic language. If we select a person at random from the voter registration list, what is the probability that he or she will be a Celtic-speaking, Caucasian highlander?

Independence

There is a special case in which the information that K is true does nothing to change our uncertainty about J (and vice versa). The earlier example with the balls in the container and the coin was such a case. Then $\Pr(J|K) = P(J)$ and, from the first of these two equations,

$$
\Pr(J \text{ and } K) = \Pr(J)\Pr(K)
$$

In this special case the two events are said to be statistically *independent* or *unassociated*.

Note that, although we have omitted E for brevity from our notation, the independence or otherwise of J and K will be determined by E. If every white ball in the urn of the previous example was marked with a T and every black ball was marked with an H, then B and T are dependent. If half of each color are marked H and half marked T, then B and T are independent. It is more correct to say that events are *independent conditional on* E.

When we discuss forensic transfer evidence later, we will encounter situations where two events are independent under one hypothesis, but dependent under another. Here is an example: in an inheritance dispute a man claims to be the brother of a deceased person. Under his hypothesis, the events that he and the deceased person have a particular DNA profile are dependent, but under the hypothesis that he is unrelated to the deceased person the two events may be taken as independent. As we will see in Chapter 4, sibs are more likely than unrelated people to have the same DNA profile.

Box 1.3: Derivation of the law of total probability
Let $S_1, S_2, \ldots, S_r$ be r mutually exclusive events. Furthermore, let them be exhaustive so that $\sum_i \Pr(S_i) = 1$. Let R be any other event. Then the events $(R \text{ and } S_1), (R \text{ and } S_2), \ldots, (R \text{ and } S_r)$ are also mutually exclusive. The event

$$(R \text{ and } S_1) \text{ or } (R \text{ and } S_2) \text{ or } \cdots \text{ or } (R \text{ and } S_r)$$

is simply R, because the S_i are exhaustive. So, from the second law

$$\Pr(R) \;=\; \Pr(R \text{ and } S_1) + \Pr(R \text{ and } S_2) + \ldots + \Pr(R \text{ and } S_r)$$

Then, by the third law

$$\Pr(R) \;=\; \sum_i \Pr(R|S_i)\,\Pr(S_i)$$

The notion of independence extends in a similar way to three or more events, as we shall see later when we combine genotype probabilities across multiple loci.

The Law of Total Probability

From the above three laws follow the entire theory of probability; no further basic laws are needed. However, there are certain standard results that are used frequently. The first of these, the law of total probability, has also been called the "law of the extension of the conversation" (see, for example, Lindley 1991). If A and B are two mutually exclusive and exhaustive events (so that $B = \bar{A}$), then for any other event H, the *law of total probability* states that

$$\Pr(H) \;=\; \Pr(H|A)\,\Pr(A) + \Pr(H|B)\,\Pr(B)$$

It is derived in a more general form from the first three laws as shown in Box 1.3, and readers who wish to do so may skip the derivation.

We use this result when we are interested in evaluating the probability of an event that depends on a number of other events that are themselves mutually exclusive. The examples we do in Chapter 6 on family trees and parentage analysis illustrate this, but a small foretaste may help. Let us imagine that we are interested in determining the probability that one of the *HUMTHO1* alleles an individual inherits is allele 8. One way of looking at this is to say that *either* the mother had the 8 allele and passed it on to the offspring *or* the father had the 8 allele and passed it on. We won't work that calculation at this stage, because it is complicated a little by the need to take account of the possibilities of

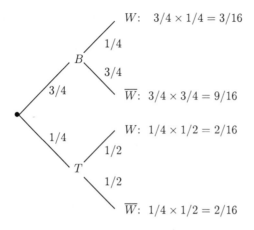

$$W: \quad 3/4 \times 1/4 = 3/16$$

$$\overline{W}: \quad 3/4 \times 3/4 = 9/16$$

$$W: \quad 1/4 \times 1/2 = 2/16$$

$$\overline{W}: \quad 1/4 \times 1/2 = 2/16$$

Figure 1.1   Tree diagram for chess tournament.

homozygosity in the parents. Instead we give an example that is based on one by Berry (1996). Fred is playing in a chess tournament. He has just won his match in the first round. In the second round he is due to play either Bernard or Tessa depending on which of those two wins their first round match. Knowing what he does about their respective abilities, he assigns 3/4 to the probability that he will have to play Bernard, $\Pr(B)$, and 1/4 to the probability that he will have to play Tessa, $\Pr(T)$. We should write these as $\Pr(B|K)$ and $\Pr(T|K)$, where K denotes Fred's knowledge of their respective abilities, but we leave the conditioning out just to simplify the notation.

What is the probability of W, the event that Fred will win? He assigns 1/4 to the probability that he will beat Bernard if he has to play him, $\Pr(W|B)$, and 1/2 to the probability that he will beat Tessa if he has to play her, $\Pr(W|T)$. So, the probability that he will win his second round match, $\Pr(W)$, is the probability of Bernard winning and Fred beating Bernard plus the probability of Tessa winning and Fred beating Tessa. That is:

$$
\begin{aligned}
\Pr(W) &= \Pr(W|B)\Pr(B) + \Pr(W|T)\Pr(T) \\
&= \left(\frac{1}{4} \times \frac{3}{4}\right) + \left(\frac{1}{2} \times \frac{1}{4}\right) \\
&= \frac{5}{16}
\end{aligned}
$$

It is useful to present this type of analysis as the tree diagram shown in Figure 1.1.

Exercise 1.3   According to the 1991 census, the New Zealand population consists of 83.47% Caucasians, 12.19% Maoris, and 4.34% Pacific Islanders. The probabilities of finding the same *YNH24* genotype as in the crime sample in the case *R. v. Ladbrook* from a Caucasian, Maori, or Pacific Islander are 0.013, 0.045, and 0.039 respectively. What is the probability of finding that genotype in a person taken at random from the whole population of New Zealand?

Odds

Before we explain another important result of probability theory, we need to explain the notion of *odds*. It is a term used in betting, where it means something slightly different from what it means in formal theory. In everyday speech, odds and probability tend to be used interchangeably; this is a bad practice because they are not the same thing at all.

If we have some event H about which the conditioning is unambiguous, then $\Pr(H)$ denotes the probability of H, and the odds in favor of H are

$$O(H) \ = \ \frac{\Pr(H)}{\Pr(\bar{H})}$$

i.e.,

$$O(H) \ = \ \frac{\Pr(H)}{1 - \Pr(H)}$$

Because $O(H)$ is the ratio of two probabilities, it can take any value between zero (when H is false) and infinity (when H is true). Let us consider some numerical examples. In the family board game, given the agreed conditions for rolling the die, we have 1/6 as our probability for the event that the die will show a score 3, and 5/6 for the probability that it will not show a score 3. The odds in favor of a 3 showing are:

$$O(3) \ = \ \frac{1/6}{5/6} = \frac{1}{5}$$

When, as here, the odds are less than one, it is customary to invert them and call them *odds against*. So, in this case, the odds are 5 to 1 against.

Now consider a fair coin toss example where we can agree that the probability of a head $\Pr(H)$ is 0.5. Then the odds in favor of a head are 0.5/0.5 = 1. Odds of one are conventionally called *evens*, no doubt as an indication that things are evenly balanced.

For a last example on converting probabilities into odds, think of the first round of the chess game described earlier. Fred's probability that Bernard will beat Tessa, $\Pr(B)$, is 3/4 so the odds in favor of Bernard winning are 3/4 divided

example, the suspect may acknowledge that the blood is his and present a credible explanation why it should have been at the crime scene for purely innocent reasons. In a rape case, for example, evidence that semen on a vaginal swab from a complainant has the same type as that of the suspect does not necessarily mean that a crime has been committed. It may be that the complainant had consented to intercourse. For this reason we will talk in terms of propositions that are more closely directed to the origin of the crime stain. In the present example, the scientist can anticipate that, if the case comes to court, the prosecution will put to the jury the following proposition:

H_p: The suspect left the crime stain.

Of course, the shorthand H_p is introduced here solely to assist us in the formal analysis that follows. We would not expect prosecuting counsel to speak in algebraic notation. We have already recognized that at court the suspect will be referred to as the defendant.

We now introduce some more notation. Let G_S and G_C denote the DNA typing results for the suspect and crime sample, respectively, and let I denote the non-DNA evidence that will be put to the court in relation to H_p. Note that in the present example $G_S = G_C$. We can now view the interpretation problem as one of updating uncertainty in the light of new information. Before the DNA evidence, the probability of H_p was conditioned by I: $P(H_p|I)$. After the DNA evidence, the probability of H_p is conditioned by G_S, G_C, and I: $\Pr(H_p|G_S, G_C, I)$.

We saw in Chapter 1 how Bayes' theorem can be used to do this. However, in the present context, we cannot proceed unless we introduce some sort of alternative proposition to H_p. In the most general case we would have a range of alternatives, but things are simplest if there is only one, and in the legal setting this is appropriate because it naturally becomes the defense proposition:

H_d: Some other person left the crime stain.

Clearly, H_p and H_d in this case are mutually exclusive and exhaustive. At this point we emphasize what we will call the *first principle of evidence interpretation*: to evaluate the uncertainty of any given proposition, it is necessary to consider at least one alternative proposition.

If we now talk in terms of odds, then our problem is one of progressing from

$$\frac{\Pr(H_p|I)}{\Pr(H_d|I)}$$

which we can call the prior odds in favor of H_p, to

$$\frac{\Pr(H_p|G_S, G_C, I)}{\Pr(H_d|G_S, G_C, I)}$$

which we call the posterior odds in favor of H_p. This can be calculated from the odds ratio form of Bayes' theorem described in Chapter 1, where now $E = (G_S, G_C)$:

$$\frac{\Pr(H_p|E, I)}{\Pr(H_d|E, I)} = \frac{\Pr(E|H_p, I)}{\Pr(E|H_d, I)} \times \frac{\Pr(H_p|I)}{\Pr(H_d|I)}$$

We consider this equation to be of central importance in forensic interpretation because it enables a clear distinction to be made between the role of the scientist and that of the juror. The jury needs to address questions of the following kind:

- What is the probability $\Pr(H_p|E, I)$ that the prosecution proposition is true given the evidence?

- What is the probability $\Pr(H_d|E, I)$ that the defense proposition is true given the evidence?

On the other hand, the scientist must address completely different kinds of questions:

- What is the probability $\Pr(E|H_p, I)$ of the DNA evidence if the prosecution proposition is true?

- What is the probability $\Pr(E|H_d, I)$ of the DNA evidence if the defense proposition is true?

We cannot emphasize this distinction enough and will return to it frequently; our *second principle of evidence interpretation* is that the scientist must always be asking questions of the kind "What is the probability of the *evidence* given the proposition?" and not "What is the probability of the proposition given the evidence?" The latter is the kind of question that falls within the domain of the court.

The posterior odds that we seek are therefore arrived at by multiplying the prior odds by a ratio of two probabilities–the likelihood ratio (LR):

$$\text{LR} = \frac{\Pr(E|H_p, I)}{\Pr(E|H_d, I)}$$

The *third principle of evidence interpretation* emerges from this analysis: The scientist must evaluate the DNA evidence, not only under the conditioning of H_p and H_d, but also under the conditioning of the non-DNA evidence, I. This is another point to which we will return frequently. But now we continue with the present example by writing out E in its two components, so the likelihood ratio is

$$\text{LR} = \frac{\Pr(G_S, G_C|H_p, I)}{\Pr(G_S, G_C|H_d, I)}$$

To take this a stage further, we need to use the third law of probability to expand the numerator and denominator of the ratio. There are two ways of doing this:

$$\text{LR} = \frac{\Pr(G_C|G_S, H_p, I)}{\Pr(G_C|G_S, H_d, I)} \times \frac{\Pr(G_S|H_p, I)}{\Pr(G_S|H_d, I)} \tag{2.1}$$

and

$$\text{LR} = \frac{\Pr(G_S|G_C, H_p, I)}{\Pr(G_S|G_C, H_d, I)} \times \frac{\Pr(G_C|H_p, I)}{\Pr(G_C|H_d, I)} \tag{2.2}$$

The first of these is called *suspect anchored* and the second is called *scene anchored*. The choice of which to use is determined by the problem at hand. In principle they should each lead to the same final answer, but one or other of them may involve probabilities that are difficult to assign given the circumstances. Interested readers are referred to Evett and Weir (1991) to see how the present problem can be solved either way. However, for the present we will use Equation 2.1.

The terms $\Pr(G_S|H_p, I)$ and $\Pr(G_S|H_d, I)$ denote the probabilities of the observation G_S for the suspect sample given either the suspect did or did not leave the crime sample. The very important point to note here is that the conditioning does not include the observation G_C for the crime sample, and whether or not the suspect left the crime sample does not provide us with any information to address our uncertainty about his genotype. So,

$$\Pr(G_S|H_p, I) = \Pr(G_S|H_d, I)$$

and the likelihood ratio simplifies to

$$\text{LR} = \frac{\Pr(G_C|G_S, H_p, I)}{\Pr(G_C|G_S, H_d, I)}$$

Now remember that $G_S = G_C$ in this particular example. If we assume that the genotype we are considering can be determined without error, then it is certain that G_C would take the value it does if H_p is the case, so $\Pr(G_C|G_S, H_p, I) = 1$ and the likelihood ratio simplifies still further to

$$\text{LR} = \frac{1}{\Pr(G_C|G_S, H_d, I)} \tag{2.3}$$

It is necessary to assign the probability of G_C given that some person other than the suspect left the crime stain. The way that we proceed from here depends on I, or, as we more commonly say, the *circumstances*. For the present, we are going to assume that the circumstances are such that knowledge of the suspect's type G_S does not influence our uncertainty about the type of the offender, given that person is not the suspect. Then, in formal terms

$$\Pr(G_C|G_S, H_d, I) \;\; = \;\; \Pr(G_C|H_d, I) \tag{2.4}$$

Because there will be cases in which it is not true, it is important to remember that this assumption has been made. One such circumstance is the situation in which if the crime stain did not come from the suspect then it came from a close relative; then knowledge of G_S certainly influences our judgment about the probability of G_C. We will be discussing such situations in later chapters. For the time being, we assume that Equation 2.4 holds. Then

$$\mathrm{LR} \;\; = \;\; \frac{1}{\Pr(G_C|H_d, I)} \tag{2.5}$$

Also, to simplify notation, for the rest of this section we let $G_C = G_S = G$.

We have now reached the point at which so much of the debate on DNA statistics starts; how do we assign a value to the denominator of this likelihood ratio? What is the probability that we would observe genotype G if some person other than the suspect left the stain? The answer depends entirely on the circumstances I. We need to address the concept of a group of people, identified according to the information I, to which the offender belongs, if that person is someone other than the suspect. This group will, in the simplest situation, be seen to be a *population* of some kind–perhaps the population of the town in which the crime was committed, or a population of a particular ethnic group identified by some element of I, such as would be the case when an eyewitness says that the person who committed the crime appeared to be Caucasian. However the population is identified, it will not normally be the case that we know everything that there is to know about all of its members. On the contrary, our information will be limited to data collected from a small portion (a *sample*) of the population. The science of using samples to make inferences about populations is called *statistics* and forms a major part of this book.

So let us assume that we have data from a sample of people we believe to be *representative* of the population to which the offender belongs, if the suspect is not the offender. At this stage, we do not discuss what we mean by "representative" other than to say that it is a matter for the judgment of the scientist in the case to decide whether he considers the data from the sample to be relevant to inference about the population suggested by I. Let us further assume, without going into the details of how we may do it, that we estimate that genotype G occurs in a proportion P of the population to which the offender credibly belongs, if he is not the suspect. Then we assign the probability P to the denominator of

Equation 2.5 and our likelihood ratio is

$$LR = \frac{1}{P}$$

If, for example, $P = 1/100$ then the likelihood ratio is 100 and the evidence could be presented in the form "The evidence is 100 times more probable if the suspect left the crime stain than if some unknown person left the stain." We discuss issues of communication later.

We wish to emphasize a couple of points here. First, in our view, the process is that the scientist *assigns* a numerical value to the denominator based on all the information available to him and his judgment of the relevance of all the different aspects of the information. Of course, we do not regard that judgment to be infallible and it goes without saying that, if the case comes to court, he will need to explain his reasons. Furthermore, he will have proceeded as he has by taking account of the circumstances of the offense as they were explained to him. If those circumstances change in any way, then it may be necessary for him to review his interpretation of the evidence. The scientist uses an estimate of the population proportion of the offender's type, G, as the probability of finding the crime sample to be of that type if it had been left by some other person.

Readers will note that we are emphasizing the personal nature of the interpretation and we stress that further by dismissing the idea that in any case there is a "right answer" when it comes to the assessment of DNA evidence. Of course, there is a right answer to the question of whether or not the suspect left the crime stain but, as we have seen, it is not within the domain of the scientist's expertise to address that question (though readers familiar with the forensic field will be aware that with other types of scientific evidence, such as handwriting and fingerprinting comparison, it has long been the scientist's function to address such questions). But as far as the likelihood ratio is concerned, scientists should not be led down the false path of believing that there is some underlying precise value that is "right." We must never lose sight of the fact that, for the denominator, we are conditioning on the idea that some unknown person left the stain. What do we mean by an "unknown person"? More importantly, what will the court determine to be the most appropriate concept for the unknown person? We have loosely invoked the concept of a population, but human populations are never completely homogeneous and can never be precisely defined. Even if we become as general as possible and say that we mean the population of the world, then do we mean the population today? or yesterday? Let's say that we mean the population of the world at the instant that the crime was committed. But do we really suggest that a female octogenarian from Beijing should be regarded a potential offender in this case? Or a 6-month-old Ghanaian? If we decide that we will consider the population of the town in which the crime was committed then are we going to ignore the possibility that a person from another town left the stain? The offender may be a visitor passing through, for example. Or, do we mean an area of the town? Are some streets more likely than others to provide refuge to the true offender? Whereas we will show in Chapter 5 that such effects are in general of little practical importance, they do mean that there is always a residual uncertainty and the concept of a "right answer" is misleading. The probability that we assign to the denominator of the likelihood ratio is ultimately a matter of judgment–informed by I. This means that the scientist must be clear in his evidence to the court of the nature of I as it appeared to him when he made his assessment. If the court has a different view of I, then this will inevitably mean that the scientist must review the interpretation of the evidence.

THE BAYESIAN MODEL

Scientists have been presenting body fluid evidence in court for decades without resorting to the Bayesian model, and it is a natural reaction at this stage to suggest that presenting the evidence in the form of a likelihood ratio is unnecessarily complicated. Why not just give the court an estimate of the frequency of the observed type?

If the case is indeed as simple as the one described above then it can be argued that the relative frequency approach is as effective as the likelihood ratio. However, as soon as any sort of complication arises–and we will meet various kinds of complications in the pages that follow–the frequency approach breaks down and can give answers that are misleading. Although we have argued that there is no "right answer" it does not follow that there are no "wrong answers": there are answers that are patently wrong–and we will meet some of them–because they lack science and logic. In cases that involve any kind of complication a Bayesian analysis is unavoidable. We claim no originality for this view, and we refer interested readers to important hallmark publications in the field. Mosteller and Wallace (1964) used Bayesian analysis to explore the issue of the authorship of "The Federalist" papers. Finkelstein and Fairley (1970) first explored transfer evidence from the perspective we have described above, and Lindley (1977) developed the ideas further in exciting ways. A comprehensive overview of evaluating scientific evidence is given in recent books by Robertson and Vignaux (1995) and Aitken (1995): the first of these is less mathematical than the second. Some description of the approach is also contained in the second NRC report (National Research Council 1996), and a very useful review was given by Friedman (1996).

Likewise, we would be remiss if we gave the impression that the Bayesian view was universally accepted–it is not. Counter views have been expressed by Tribe (1971) and Kind (1994). However, among serious students of inference reports. In practice, samples are not collected anew for every crime, and we shall see in Chapter 5 how this is accommodated. We will see there that frequencies of DNA profiles may differ between populations, especially when people in these populations have different racial backgrounds. Especially under hypothesis H_d, the racial background of the suspect does not define the population.

In 1991, a Mr. Passino was on trial for homicide in Vermont. His defense established that his paternal grandparents were Italian, his maternal grandfather was Native American and his maternal grandmother was half French and half Native American. On this basis, the defense was able to have DNA profile calculations ruled inadmissible because they were not based on a sample of people with the same racial heritage as Mr. Passino. We pointed out (Weir and Evett 1992) the lack of logic since the defense hypothesis was that the crime sample was not from Mr. Passino, and therefore his racial background was immaterial. We agree with Lewontin (1993), who noted that the circumstances of the crime suggested that the offender may have been a member of the Abnaki tribe of Native Americans, although this still does not require account to be taken of Mr. Passino's pedigree (Weir and Evett 1993).

THE TWO-TRACE PROBLEM

We now consider a case that is a little more complicated than the first. In this case, examination of the crime scene reveals stains of two different types. Assume that the investigator is justified in inferring that the stains were left during commission of the offense, implying that there were two offenders with different blood types. Assume, further, that information received by the investigator leads to the detention of a single suspect, who provides a blood sample. DNA typing yields the genotypes G_1, G_2 from the two crime samples and G_S from the suspect sample. The suspect's type matches that of one of the stains, $G_S = G_1$, but not the other.

In this case we can visualize the two propositions as

H_p: The suspect was one of the two men who left the crime stains.
H_d: Two unknown men left the crime stains.

As before, we seek to evaluate

$$LR = \frac{Pr(E|H_p, I)}{Pr(E|H_d, I)}$$

where now $E = (G_S, G_1, G_2)$. Making assumptions similar to those embodied in Equation 2.3, this can be shown to be

$$LR = \frac{Pr(G_1, G_2|G_S, H_p, I)}{Pr(G_1, G_2|G_S, H_d, I)}$$

Remember that $G_S = G_1$.

Numerator. This is evaluated by posing the question "If the suspect and some unknown man left the crime stains, what is the probability that one of the stains would be G_1 and the other G_2?" If we repeat the assumption that the types are determined without error, then with probability one, the suspect would leave a stain that would give genotype G_1. If we are told that the proportion of people, in what we judge to be the most relevant population, who would give observation G_2 is P_2, then the answer to the question is $1 \times P_2$, i.e.,

$$\Pr(G_1, G_2 | G_S, H_p, I) \quad = \quad P_2$$

Denominator. This is evaluated by posing the question "What is the probability that two unknown people would leave stains giving observations G_1 and G_2?" Given that the proportions of the two types in the relevant population are P_1 and P_2 respectively, then it is tempting to reply, making an obvious assumption, that the answer is $P_1 \times P_2$. However, a moment's reflection is needed. Think, for a moment about tossing two coins–say a dime and a quarter–and ask yourself the probability of two heads being the result. The answer is straightforward: the dime will show a head with probability 0.5, and the quarter will also show a head with probability 0.5. If we specify that the tossing has been done properly, then these two events are independent and the probability of two heads is their product: 0.25. What now if you are asked the probability that the result of tossing the two coins is one head and one tail? There are two ways in which this can happen:

- The dime shows a head and the quarter shows a tail.

- The dime shows a tail and the quarter shows a head.

So the answer, that is another example of applying the law of total probability, is $(0.5 \times 0.5) + (0.5 \times 0.5) = 0.5$.

By exactly the same reasoning, the probability that two unknown men would give us observations G_1 and G_2 is $2P_1 P_2$. You could also visualize this by imagining the two men walking in through your door: the result you seek occurs if the first man is G_1 and the second man is G_2 or if the first man is G_2 and the second man is G_1. So

$$\Pr(G_1, G_2 | G_S, H_d, I) \quad = \quad 2P_1 P_2$$

It follows that the likelihood ratio is

$$\text{LR} \quad = \quad \frac{1}{2P_1}$$

Note that the likelihood ratio is half what it would have been if there had only been one stain. That the likelihood ratio is less in the second case is intuitively reasonable.

There is an interesting feature of this result. Note that if the proportion P_1 is greater than 0.5, the likelihood ratio is less than one. Even though there is evidence that appears to confirm the prosecution hypothesis, it actually supports the defense hypothesis. In Chapter 9, when we discuss case reporting, we will explain the unsatisfactory nature of phrases such as "consistent with." A classical forensic approach to this sort of case might have been to report a match and say something like "the evidence is consistent with the presence of the suspect's blood." We now see that this approach does not offer a balanced interpretation of the evidence.

A likelihood ratio less than one when there is a match between a suspect and the crime sample may appear counter-intuitive at first sight, but its validity can be illustrated by a simple example. Imagine that two marks have been made on a white board, one by a red pen and one by a green pen. The two pens have been dropped into an opaque container that contains 98 other pens of the same shape and size. A person is told there are now 60 red pens and 40 pens that are not red in the container, and is asked to reach into the container, withdraw a pen and speculate whether or not it was one of the two pens used to mark the white board. While the person has the pen in his hand, but before he has withdrawn his hand from the container and observed the pen, it would be reasonable for him to assign a probability of 1/50 that it was one of the two pens used. However, if he finds that the chosen pen is red, the probability that it was one of those used is 1/60. Even though the pen has the right color to have marked the white board, and so "matches," the new information (the color of the pen drawn) has actually reduced the probability that it is one of the two being sought.

TRANSFER FROM THE SCENE

In the previous two cases, the offender or offenders left evidence at the crime scene. In this case we consider interpretation when it is possible that the offender inadvertently took evidence from the crime scene.

We imagine a crime in which a victim has bled profusely after being stabbed. As a result of an investigation, a suspect is arrested and his outer clothing taken for scientific examination. Blood staining on the clothing is found to be of the same genotype G as the victim's blood. The suspect himself has a different genotype. The important difference between this and the preceding cases is that not only does the genotype match have evidential value, but also the very presence of blood staining on the suspect's clothing will need to be taken into account. We use $E = (E_1, E_2)$ to denote the two aspects of the bloodstain evidence:

- E_1: There is blood staining on the suspect's clothing.

- E_2: The blood staining on the suspect's clothing has genotype G.

For generality, let G_V and G_S denote the genotypes of the victim and suspect, respectively, remembering that $G_V = G$ and $G_S \neq G$. We consider two propositions:

H_p: The suspect is the person who stabbed the victim.
H_d: The suspect is not the person who stabbed the victim.

We will assume that the circumstances I make it clear that there was only one person involved in the assault on the victim, and there are no other mechanisms by which the victim's blood could have been transferred to the suspect. Then the likelihood ratio is

$$\text{LR} = \frac{\Pr(E, G_V, G_S | H_p, I)}{\Pr(E, G_V, G_S | H_d, I)}$$

The first stages of simplification can be done using the multiplication law as follows:

$$\text{LR} = \frac{\Pr(E, G_V | G_S, H_p, I)}{\Pr(E, G_V | G_S, H_d, I)} \times \frac{\Pr(G_S | H_p, I)}{\Pr(G_S | H_d, I)}$$

There is no information in H_p or H_d that would influence the genotype of the suspect, so the second ratio is one and we move to the next stage.

$$\text{LR} = \frac{\Pr(E | G_V, G_S, H_p, I)}{\Pr(E | G_V, G_S, H_d, I)} \times \frac{\Pr(G_V | G_S, H_p, I)}{\Pr(G_V | G_S, H_d, I)}$$

By similar reasoning to that in the previous stage we argue that the second ratio is one. Also, if H_d is true then E is independent of G_V. Then

$$\text{LR} = \frac{\Pr(E | G_V, G_S, H_p, I)}{\Pr(E | G_S, H_d, I)}$$

We will now consider the numerator and denominator in turn.

Numerator. To assist with the numerator, we introduce a new proposition T that blood was transferred from the victim to the assailant's clothing. The complementary event $\bar{T}$ is that blood was not transferred. Then we use the law of total probability:

$$\begin{aligned}
\Pr(E | G_V, G_S, H_p, I) = \ & \Pr(E | T, G_V, G_S, H_p, I) \Pr(T | G_V, G_S, H_p, I) \\
& + \Pr(E | \bar{T}, G_V, G_S, H_p, I) \Pr(\bar{T} | G_V, G_S, H_p, I)
\end{aligned}$$

ences about populations from random samples. Yet we have seen that in the forensic context, we will generally be dealing not with random, but with convenience, samples. Does this matter? The first response to that question is that every case must be treated according to the circumstances within which it has occurred, and the next response is that it is always a matter of judgment. The theory of statistics, upon which most of this book is based, operates within a framework of assumptions, but it needs to be applied to real-life problems. The forensic scientist needs to judge whether the assumptions appear reasonable in the individual case. The scientist should consider the literature, but must also ask if there is any reason to believe that knowledge of a person's sex, age, socioeconomic status, political persuasion, or tendency to criminality would in any way provide information to address the uncertainty about his genotype. In the last analysis, the scientist must also convince a court of the reasonableness of his or her inference within the circumstances as they are presented in evidence. This cause may be helped by statements in the 1996 report of the United States National Research Council (National Research Council 1996) that the loci used for identification are unlikely to be correlated with traits associated with different subsets of the population, and that frequencies of alleles at these loci do not differ very much among different subpopulations of geographic areas.

Although the term "random sample" is being applied to genotypes, not to people, we tend to ignore this distinction. Should the scientist consider his convenience sample to be a random sample? To some extent that depends on the typing system that he is considering. If it were feasible for him to identify the most relevant subset of the population of Gotham City, pick a sample at random from this subset, find the selected individuals, and persuade them to provide body fluid samples, would this truly random sample look much different from his current convenience sample? Certainly it would not be precisely the same, but would the differences have any practical effect? We believe that the scientist should be considered competent to address questions of this nature and to make judgments about what are and are not "practical effects." If the scientist is satisfied in this regard then, we maintain that he or she can proceed as though the convenience sample were a random sample.

Using our sample of individuals we will attempt to draw inferences about a population in order to assign a value to the denominator of the likelihood ratio. This value will be related, often in a very simple way, to the proportion of people in the population who have genotype G. But, as we have seen, we will not know that proportion, and indeed it is almost always unknowable, for the reasons we have sketched in previous paragraphs. We will use our sample, regarded as effectively random, to estimate the proportion.

Table 3.1   List of 50 *FES* genotypes.

10,10	10,12	11,12	12,12	10,11
10,13	11,12	10,11	11,12	10,11
11,12	8,12	10,10	10,11	10,11
10,11	10,11	10,12	10,11	11,11
11,11	10,13	12,13	10,12	11,12
10,11	11,11	10,13	11,12	11,11
11,12	10,11	11,12	11,12	11,12
10,11	11,12	10,11	11,11	11,11
10,11	10,10	10,11	11,11	10,12
10,10	10,11	11,11	8,11	10,12

Example

We will now make our example a little more concrete by saying that genotype G is 11,12 at the *HUMFES/FPS* locus, which we will call *FES* for short. Our sample consists of 50 Caucasians whose genotypes are in Table 3.1. The integers 8, 10, 11, 12, and 13 indicate alleles, and the genotype of each individual is given by the two alleles the individual possesses at this locus. One allele has come from each of the individual's parents.

We can summarize the data conveniently as the counts for all the 15 genotypes possible with 5 alleles. If previous samples had indicated the possibility of allele 9, then we may want to include those genotypes that include that allele even though they are absent for this sample. The genotype counts are displayed in the body of Table 3.2 for the alleles specified in the row and column margins.

Table 3.2   Genotype counts for the *FES* sample.

Allele	Genotype counts					
8	0					
9	0	0				
10	0	0	4			
11	1	0	15	8		
12	1	0	5	11	1	
13	0	0	3	0	1	0
Allele	8	9	10	11	12	13

In our sample 11 people out of 50 have the genotype 11,12. Is the figure $11/50 = 0.22$ the one that we should use to assign a value to the denominator of the likelihood ratio? The short answer is that it is the best information we have, but we must recognize that it is not the same thing as the proportion P in the population. Not only, as we have seen, is the notion of population rather vague, but also a sample can provide only an *estimate* of the population proportion. We need to understand the properties of estimators: How good an estimator of a population proportion is a sample proportion? To understand such issues we need to study some more theory, and we next introduce the *binomial distribution*.

BINOMIAL DISTRIBUTION

An Urn Model: Two Kinds of Balls

Equal proportions of the two kinds of ball. We will learn about the binomial distribution by returning to the model in Chapter 1 of a large urn that contains a number of balls. All of the balls are indistinguishable from each other in size and shape. They differ in color, however, and for the binomial ("two names") we imagine that there are two colors: white and black. We will be considering conceptual experiments that involve drawing one ball at a time in such a way that our drawing process is completely insensitive to the color of the ball. We can do this by not looking inside the urn and by giving the balls a good stir between draws. The color we end up with after a single draw is uncertain, though we know that the greater the number of balls of a given color, the more likely we are to end up with that color. Thus, to speculate about the outcome of a draw we would like to know the proportions of the two colors in the urn. The proportions of the two colors remaining in the urn would change if we did not return the ball we drew to the urn. For this discussion we are going to return each ball we draw after we have noted its color, and this is called *sampling with replacement*.

If, before drawing a ball, we are told that half of the balls in the urn are black and half are white, then the following statement should seem reasonable: "The probability that the ball will be black is 0.5."

The first experiment we consider consists of drawing one ball, noting its color and replacing it, and then drawing a second ball, and also noting its color. If B and W denote the colors, then there are four possible outcomes: BB, BW, WB, or WW. Each of these has a probability $0.5 \times 0.5 = 0.25$ because it seems reasonable to consider successive drawings from the urn as independent of each other. Suppose we are interested only in the number of each color, and we are indifferent to the order in which they appear. Then we can write out the three outcomes (0,1, or 2 black balls) in tabular form:

Number of black balls	Possible drawings	Number of ways	Probability
0	WW	1	0.25
1	BW, WB	2	0.50
2	BB	1	0.25

The third column reminds us that there are two ways (BW, WB–each with probability 0.25) in which we can get one black ball, but only one way in which we can get none and only one way of getting two.

We can construct a similar table for the number of black balls seen when three are drawn, noting that the probability for any particular sequence (e.g. WWB) is now $0.5 \times 0.5 \times 0.5 = (0.5)^3$

Number of black balls	Number of ways	Probability
0	1	0.125
1	3	0.375
2	3	0.375
3	1	0.125

If we wish to extend this sort of analysis to more and more balls, then we can use *Pascal's triangle* to work out the numbers of ways of getting a given number of blacks. The first four rows of this triangle are shown in Table 3.3, and each number in the triangle is seen to be the sum of the two numbers to the left and right of it in the preceding line. Completing the next line, corresponding to drawing six balls, should be a simple exercise for the reader.

Let's look at the fourth line in a little more detail. This is for the case in which we are drawing five balls, and we see (Table 3.4) that there are 10 ways of drawing two black balls, for example.

Pascal's triangle can be cumbersome when it comes to making such calculations for drawing large numbers of balls. To derive a general formula, consider

Table 3.3   Pascal's triangle.

		1	2	1		
	1	3	3	1		
1	4	6	4	1		
1	5	10	10	5	1	

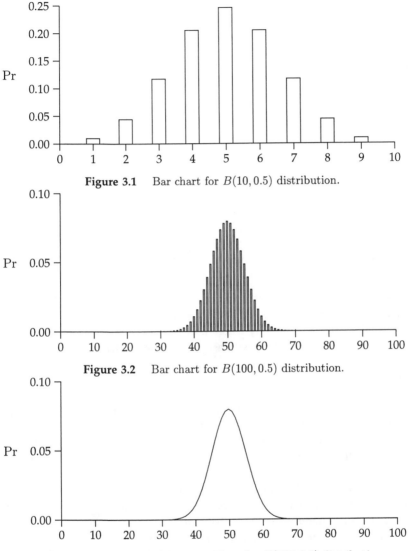

Figure 3.1   Bar chart for $B(10, 0.5)$ distribution.

Figure 3.2   Bar chart for $B(100, 0.5)$ distribution.

Figure 3.3   Join of midpoints of bars for $B(100, 0.5)$ distribution.

areas under the pdf between the limits for the range, and the total area under a pdf is equal to one (mathematically, the integral of the pdf over its range is one).

Equation 3.2 is not convenient for giving numerical values, and we use tables of values instead. Although there are infinitely many different normal distri-

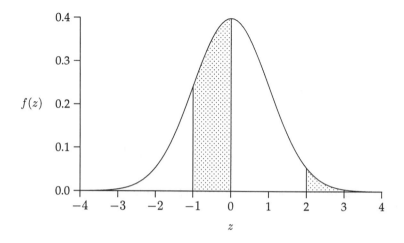

Figure 3.4   Shaded areas are probabilities of $-1 < z < 0$ and $2 < z < 3$.

butions, corresponding to all the possible means and variances, they can all be rescaled into the **standard normal** that has a mean of zero and a variance of one, $N(0, 1)$. Values x of any quantity whose uncertainty is described by a $N(\mu, \sigma^2)$ distribution can be rescaled to z, with an $N(0, 1)$ distribution, by

$$z \;=\; \frac{x - \mu}{\sigma} \sim N(0, 1) \tag{3.4}$$

As we will soon see, the standard normal distribution has the useful property that about 95% of the distribution lies within two standard deviations of its mean. The distribution is described in Appendix Table A.1. Entries in the body of that table provide the probability of z being greater than the value specified in the margins. The table shows that the probability of z being greater than 1.00 is 0.1587. To find the probability of z being between 2.00 and 3.00 we use the table to find the probability (0.0228) of z being greater than 2.00 and subtract from that the probability (0.0013) of z being greater than 3.00. The difference of 0.0215 is the probability for the range 2.00 to 3.00 (Figure 3.4).

Symmetry of the standard normal pdf means that the probability of z being less than -1.5, for example, is the same as that of z being greater than $+1.5$ (0.0668). Symmetry also means that the probability of z being between zero and -1.00 is the same as for the range zero to $+1.00$, and this value is the difference between the probability (0.5000) of z being greater than zero, and the probability (0.1587) of z being greater than 1.00. This difference is 0.3413 (Figure 3.4).

The most commonly used z value is 1.96. We see in Table A.1 that the area to the right of this value is 0.0250, and therefore the area to the left of 1.96 is 0.9750. This means that 1.96 is the **97.5th percentile** of the standard normal distribution. Symmetry means that a total of 5% of the area under the standard normal curve lies outside the range ± 1.96. In other words, there is a probability of 95% that a random quantity with the standard normal distribution will have a value between ± 1.96. Equation 3.4 extends this result to mean that there is 95% probability that a random quantity with any normal distribution will have a value within 1.96 standard deviations of the mean for that distribution (i.e., approximately within two standard deviations of the mean).

The standard normal approximation to the binomial variable is obtained from

$$z \;=\; \frac{x - np}{\sqrt{np(1-p)}} \sim N(0,1)$$

and this allows probability statements to be made about counts x of a discrete binomial quantity by making use of the continuous normal distribution. For large values of n, it is much easier to refer to tables like Table A.1 than it is to evaluate $n!$.

As an example, consider sample allele proportions for the ABO blood group system. For a sample of size 16 alleles from a population in which allele A has proportion 0.50, the probabilities of x A alleles ($x = 0, 1, \ldots, 16$) are given by the $B(16, 0.5)$ distribution and are shown in Table 3.6. These values show that the probability of obtaining 6 or fewer As is 0.2272. The corresponding normal approximation to the probability is found by calculating the values of z corresponding to the range $x \leq 6$. These values are

$$
\begin{aligned}
z \;&=\; \frac{6 - np}{\sqrt{np(1-p)}} \\
&\leq\; \frac{6 - 8}{\sqrt{16 \times 0.5 \times 0.5}} \\
&=\; -1.0
\end{aligned}
$$

Table A.1 shows that the area under the standard normal curve to the right of 1.0 is 0.1587, so the area to the left of -1.0 is also 0.1587. The normal-approximation probability of 0.1587 is somewhat different from the exact value of 0.2272. However, the quality of the normal approximation is improved with a continuity correction that reduces the magnitude of the numerator of z by 0.5. (Positive values of the numerator are reduced by 0.5 and negative values are increased by 0.5.) This makes $z = -0.75$, and the tabulated probability is then 0.2266, which is very close to the exact value.

The quality of the normal approximation to the binomial diminishes as p deviates from 0.5. For $p = 0.20$, the population proportion of the B allele,

Table 3.6   Probabilities for $B(16, p)$ distribution.

x	$p = 0.50$		$p = 0.20$	
	$\Pr(x\vert p)$	$\sum_{y=0}^{x}\Pr(y\vert p)$	$\Pr(x\vert p)$	$\sum_{y=0}^{x}\Pr(y\vert p)$
0	0.0000	0.0000	0.0281	0.0281
1	0.0002	0.0003	0.1126	0.1407
2	0.0018	0.0021	0.2111	0.3518
3	0.0085	0.0106	0.2463	0.5981
4	0.0278	0.0384	0.2001	0.7982
5	0.0667	0.1051	0.1201	0.9183
6	0.1222	0.2272	0.0550	0.9733
7	0.1746	0.4018	0.0197	0.9930
8	0.1964	0.5982	0.0055	0.9985
9	0.1746	0.7728	0.0012	0.9998
10	0.1222	0.8949	0.0002	1.0000
11	0.0667	0.9616	0.0000	1.0000
12	0.0278	0.9894	0.0000	1.0000
13	0.0085	0.9979	0.0000	1.0000
14	0.0018	0.9997	0.0000	1.0000
15	0.0002	1.0000	0.0000	1.0000
16	0.0000	1.0000	0.0000	1.0000

Table 3.6 shows the binomial probability of 6 or fewer Bs is 0.9733, and the corresponding z value is $z = +1.72$ (after applying the continuity correction). The tabulated normal probability is 0.9573.

INDUCTION

So far, we have considered only problems of **deduction**. Given a particular distribution we can make deductive statements about the outcome of a given experiment. We may not be able to predict the outcome with certainty but we can calculate probabilities of the various outcomes using mathematical methods that are, in principle at least, straightforward and noncontroversial. We now turn to a more difficult class of problem. Given the outcome of an experiment, how do we make inferences about the underlying distribution? For example: from a sample of n people, if we have found x occurrences of genotype G, what can we say about the proportion of genotype G in the population that has been sampled?

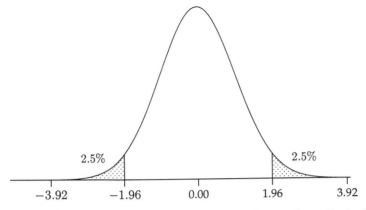

Figure 3.6   Normal distribution, showing extreme 5% of values.

an approximate 95% confidence interval. The approximation is good for large values of n; otherwise we need to use the t-distribution instead of the normal distribution. This is explained in statistics textbooks, and we will avoid the issue.

Earlier in the chapter we showed that the normal distribution can provide a good approximation to the binomial. If the probability of an event is p, the probability of x events occurring in n trials is given exactly by the binomial distribution $B(n, p)$ and approximate values of the probability can be found from the normal distribution $N(\mu, \sigma^2)$ where $\mu = np, \sigma^2 = np(1 - p)$. For the sample proportion $\hat{p} = x/n$, the normal distribution is $N(p, p(1 - p)/n)$. Therefore, the 95% confidence interval for p, when $\hat{p} = x/n$ is the sample proportion, is

$$\hat{p} \pm 1.96\sqrt{\hat{p}(1 - \hat{p})/n} \tag{3.7}$$

What about our public opinion survey? In that example, $\hat{p} = 0.47$ and $n = 1000$. Substituting these into Equation 3.7 does indeed give a confidence interval of 0.47 ± 0.03, as suggested above.

It is important to avoid the common misconception that a confidence interval provides a probability statement about the unknown quantity. A 95% confidence interval of $(0.22, 0.28)$ for an allele proportion, for example, should not be interpreted as meaning that there is 0.95 probability that the proportion lies in the interval. The correct interpretation is that, "in the long run," 95% of such confidence intervals will contain the population proportion. The phrase "in the long run" means that we cannot talk about this particular instance; it means that if we follow the same procedure in a large number of similar situations then the percentage of occasions in which the interval contains the correct proportion is 95%.

Exercise 3.11   For the FES data in Table 3.2, calculate a 95% confidence limit for the proportion of: (a) Homozygotes of type 11,11; (b) Heterozygotes of type 12,13.

BAYESIAN ESTIMATION

We now consider estimating the proportion of FES 11,12 genotypes in Gotham City from a Bayesian perspective. We have seen that the method of maximum likelihood gives an estimate of $11/50 = 0.22$.

As in the frequentist approach, we wish to make inferences about the unknown quantity p using the sample data. One of the features of the Bayesian approach is the recognition that we may well have some prior information about p. So let us start by reflecting on our knowledge before the sample had been collected. Let us first imagine that we had absolutely no knowledge of p. This would be a rather unrealistic state of affairs because there would have been some previous work to demonstrate that the FES locus was polymorphic, and that would have shed at least some light on the value, but let us discount that knowledge for the time being and imagine that we are unable to favor any particular value for p. One way of representing our state of knowledge employs the *uniform distribution* that has the continuous pdf shown in Figure 3.7. We use the notation $\pi(p)$ for the pdf of a parameter.

Note that the value of the pdf $\pi(p)$ in Figure 3.7 is one for all values of p and this satisfies the condition that the area under the graph between $p = 0$ and $p = 1$ is one. Assigning a probability density to p is very different from the frequentist view, which does not permit probability statements about unknown parameters such as p. Using $\pi(p)$ to describe our uncertainty about the unknown parameter p is central to the Bayesian view of the problem.

Once we have a sample, we have to ask how that changes our knowledge of the pdf of p, and the solution is found in the application of Bayes' theorem. In Chapter 1, we saw how the theorem was used for weighing two hypotheses against each other:

Posterior odds   =   Likelihood ratio $\times$ Prior odds

If we are considering the probability of one of several hypotheses, then the last equation in Box 1.4 can be expressed as

Posterior probability $\propto$ Likelihood $\times$ Prior probability   (3.8)

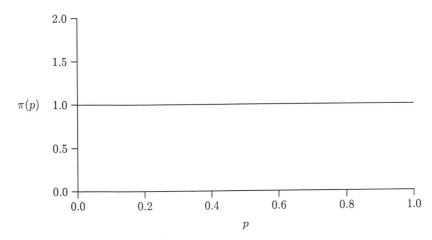

Figure 3.7   Uniform distribution.

When dealing with probability density functions we are essentially dealing with an infinite number of hypotheses, and Bayes' theorem works in the same way.

Let $\pi(p)$ denote the prior pdf for p, and let $\pi(p|x)$ denote the posterior pdf. Also, let $\Pr(x|p)$ denote the probability of the data x given p. Then, for any value of p, Bayes' theorem leads to

$$\pi(p|x) \quad \propto \quad \Pr(x|p)\pi(p) \tag{3.9}$$

The term $\Pr(x|p)$, apart from a constant of proportionality, is the same as the likelihood $L(p|x)$ we met in the section on maximum likelihood estimation. It is defined by Equation 3.5 for the specific value $x = 11$, and was shown in Figure 3.5 for that x value. Now, however, instead of taking the p value that maximizes the curve as the estimator of p, we will take the additional, and important, step of combining the curve with the prior distribution as shown in Equation 3.9. The basis of the combination is simple multiplication of the two functions for every value of p. We do not describe the additional step of integration needed to provide the constant of proportionality in Equation 3.9 that ensures that the posterior pdf has an area underneath it of one. The posterior distribution is shown in Figure 3.8. In Box 3.3 we consider a more general class of prior distributions– the *Beta* distribution.

A few features of Figure 3.8 are worth discussing. First, note that the vertical scale runs from zero to 10 and the maximum value is about 7: this is a consequence of the requirement that the total area under of the curve be one. Note

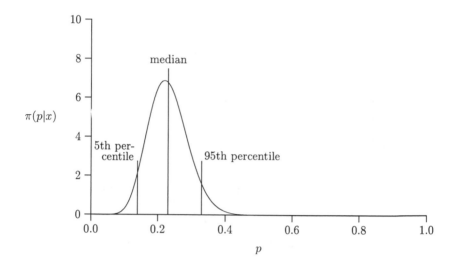

Figure 3.8   Posterior distribution for p, given $x = 11$ and the uniform prior in Figure 3.7.

that this was not the case with the graph of the likelihood function–which is *not* a pdf. Note also that it is useful to summarize the curve by means of percentiles, which are values that divide distributions into hundreths. The *median* divides a distribution in half, and 90% of a distribution lies between the *5th* and *95th percentiles*.

So our posterior knowledge, based on the sample of 50, and assuming complete prior ignorance for p, can be summarized by a median value of 0.23 and 5th and 95th percentiles of 0.14 and 0.33 respectively. It would be logically legitimate for us to say that there is a 0.9 probability that p lies between 0.14 and 0.33, but we should be cautious. All of our calculations have been based on assumptions. We assumed that our sample of 50 was representative of the population relevant to answering the question posed by the denominator of the likelihood ratio; and we have also assumed that the conditions for binomial sampling have been satisfied. So we could make the probability statement about p if we wished, but we should also make the conditioning clear. The problem with this, particularly in the forensic setting, is that while we can quantify things within the framework of our assumptions, it is generally not possible for us to quantify the effects of

Box 3.3: General Beta prior

Instead of a uniform prior, we can consider the Beta distribution $Be(\alpha, \beta)$ for p. The pdf is

$$f(p) = \frac{\Gamma(\alpha+\beta)}{\Gamma(\alpha)\Gamma(\beta)}p^{\alpha-1}(1-p)^{\beta-1}, \ 0 \le p \le 1$$

The *gamma function* $\Gamma(x)$ generally has to be evaluated numerically, but if x is an integer, $\Gamma(x) = (x-1)!$. When $\alpha = \beta = 1$, the Beta reduces to the uniform distribution.

Multiplying the Beta by the binomial distribution $B(2n, p)$ for a sample of $2n$ alleles, and canceling the terms not involving p gives

$$\pi(p|x) = \frac{p^{\alpha+x-1}(1-p)^{\beta+2n-x-1}}{\int_0^1 p^{\alpha+x-1}(1-p)^{\beta+2n-x-1}dp}$$

$$= \frac{\Gamma(\alpha+\beta+2n)}{\Gamma(\alpha+x)\Gamma(\beta+2n-x)}p^{\alpha+x-1}(1-p)^{\beta+2n-x-1}$$

The posterior distribution is also a Beta distribution, but with parameters modified by the data. In other words, the Beta is a *conjugate distribution* for the binomial. Although the whole posterior distribution is now available for p, it may be convenient to take a single feature of this distribution to serve as a Bayesian estimator of p. For example, the mean of this distribution is

$$\mathcal{E}(p|x) = \frac{\alpha+x}{\alpha+\beta+2n}$$

and the maximum of the posterior pdf is at

$$\max \pi(p|x) = \frac{\alpha+x-1}{\alpha+\beta+2n-2}$$

those assumptions and their reliability in real-world situations.

The reader may, at this stage, be rather discouraged by the size of the 90% probability interval for p but should bear in mind that we have assumed complete ignorance for our prior, and this is rarely the case in practice. We will now illustrate how this may be improved.

We may believe that the Caucasians in Gotham City are different from Caucasians elsewhere in the world, but the extensive data collected by Budowle and

Table 3.7   Frequencies of genotypes in the sample of 423 British Caucasians.

	8	9	10	11	12	13
8	0					
9	0	0				
10	5	0	37			
11	3	3	120	66		
12	3	0	54	66	26	
13	0	0	11	17	12	0

Monson (1993) shows that variation at restriction fragment length polymorphism (RFLP) loci is small, and studies of short tandem repeat (STR) data, though less extensive at the time of writing, suggest a similar picture. So in the light of such work, we may agree that Caucasian data collected by other workers is relevant to our problem of determining the proportion of Gotham City Caucasians who are genotype G. To illustrate how this can effect our evaluation, we take data for the *FES* locus from 423 British Caucasians collected by the Forensic Science Service as reported by Gill and Evett (1995). The data are displayed in Table 3.7.

We see that there were 66 observations of genotype 11,12 in the sample of 423. Note that the maximum likelihood estimator from this sample for p is therefore 0.16, rather than the 0.22 observed from the Gotham City sample. We could, if we thought it appropriate, use this sample to form our prior distribution for p, which would look like the curve in Figure 3.9. Note that both the horizontal and vertical scales have changed, and this marks the greater knowledge that the sample of 423 brings. The likelihood function for our data ($x = 11$) has not changed, and the posterior distribution is calculated as before. It is shown in Figure 3.10.

The posterior distribution is scarcely sharper than the prior, because the new sample is a lot smaller than the original sample, but its peak is slightly to the right, reflecting the fact that the genotype is more frequent in the Gotham City sample. The posterior median is now 0.164, and the 90% probability range is from 0.137 to 0.193.

Would this be a legitimate procedure to follow in the Gotham City example? Well, here we come once again to unquantified issues of the scientist's judgment. Is it right that the British Caucasian data should dominate the median frequency so powerfully? There is no simple answer to this, and here we must once again recognize the limit of the powers of statistics. Statistics enables quantifiable statements to be made only within a framework of assumptions; in any given situation it is the role of the scientist to make qualitative judgments.

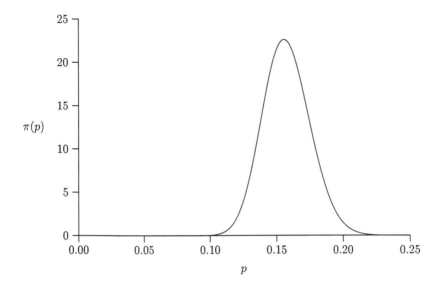

Figure 3.9   Prior distribution for p, based on a previous sample of 423 Caucasians in which 66 were of genotype 11,12.

Summary of Estimation

So far in this chapter we have concerned ourselves with problems of estimation, using as an example that of drawing inferences about the proportion of people in Gotham City who are of a certain genotype. We have seen that the method of maximum likelihood gives a single point estimate. The frequentist view also leads to the notion of a confidence interval that, in the long run, will contain the unknown value with a specified probability. We also saw that the Bayesian view of the estimation problem is to give a probability distribution for the proportion of interest. We now turn to a related issue: hypothesis testing.

TESTING HYPOTHESES

The Bayesian approach is directed to establishing a posterior probability for a hypothesis, or a posterior probability distribution for an unknown quantity. The frequentist approach is quite different in that it does not permit probability statements about hypotheses. In the same way, it is not permissible to establish a probability distribution for an unknown quantity. Instead, the frequentist ap-

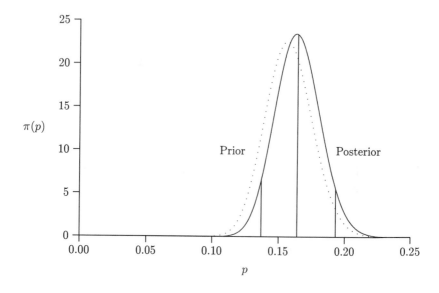

Figure 3.10   Posterior distribution (solid line) for p, given the prior distribution (dotted line) in Figure 3.9.

proach is directed toward significance testing, the essential nature of which can be illustrated by the "goodness-of-fit" test.

Goodness-of-Fit Test

We illustrate goodness-of-fit testing by means of a simple example involving roulette wheels. Apart from the zero and double zero, roulette wheels have 36 numbers: 18 red and 18 black, and we will base our discussion on a wheel with just the 36 red or black numbers. A gambler suspects that the roulette wheel in his local casino is being operated in an unfair manner, so that red numbers come up more frequently than black numbers.

If 20 consecutive spins of the wheel result in 16 reds, is this evidence that the roulette wheel is unfair? The classical approach to this question is to set up a null hypothesis that is to be tested by the data. In this case, the null hypothesis is that the wheel is fair. For goodness-of-fit testing, the first step is to calculate the numbers of reds and blacks that would be expected if the null hypothesis were true; in this case these numbers are both 10. Next, a test statistic is devised.

For categorical data, the simplest test is the ***chi-square goodness-of-fit*** test. This procedure compares observed and expected *counts* in *all* categories, squares the differences to remove sign, and divides by expected numbers to give greatest weight to largest proportional differences. The test statistic is written as X^2:

$$X^2 = \sum_{\text{categories}} \frac{(\text{Observed} - \text{Expected})^2}{\text{Expected}}$$

When the null hypothesis is true, this statistic has a chi-square distribution that, in this example, has one ***degree of freedom (df)***. The degrees of freedom can be determined as the number of expected counts that can be assigned without reference to other expected counts. In this case, the expected number of reds could be set to any number from zero to 20, but then the expected number of blacks is specified. The shape of the 1 df chi-square distribution is shown in Figure 3.11, where $f(X^2)$ is the probability density for the X^2 statistic. The shaded area indicates the probability of obtaining that value of X^2, *or a greater value*, when the null hypothesis is true. These areas are displayed in Table A.2, and show that the value 3.84 delimits the largest 5% of the distribution (it is not a coincidence that 3.84 is the square of 1.96, because the square of a quantity with a standard normal distribution has a chi-square distribution with 1 df). A X^2 value greater than 3.84 would occur with probability less than 0.05 if the null hypothesis is true, so such values are used to reject the null hypothesis at the 5% ***significance level*** or with a 5% ***P-value***. Note that X^2 will be large if there are many more reds than expected, or if there are many less reds than expected. Large departures from expectation in both directions lead to rejection, and the test procedure is said to be ***two-tailed***. The term "two-tailed" refers to the hypotheses–in this case, the alternative to the hypothesis being tested has two regions: either more or less reds than expected. The term does not refer to the single tail of the chi-square distribution shown in Figure 3.11.

Does the observation of 16 red numbers in 20 spins of a roulette wheel support the hypothesis that the wheel is fair? The observed and expected counts in all categories (reds and blacks) are shown in Table 3.8, along with the goodness-of-fit test statistic calculations. The value of $X^2 = 7.2$ is very large compared to 3.84, and Table A.2 shows that it belongs to the least probable set of large values if the wheel is fair, where "least probable" means that set having a probability between 0.005 and 0.01. The analysis would be reported as $X^2 = 7.2(P < 0.01)$.

The goodness-of-fit test applies to more than two categories. The 36 numbers on a roulette wheel are divided into three dozens: première, milieu, and dernière. A number from each of these three is equally likely when a fair wheel is spun. What can be said about a wheel that in 20 spins gave 5, 7, and 8 numbers in the three dozens? The calculations are set out in Table 3.9. As two of the expected

LIVERPOOL
JOHN MOORES UNIVERSITY
AVRIL ROBARTS LRC
TEL. 0151 231 4022

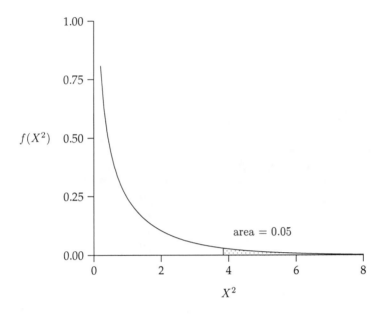

Figure 3.11   Chi-square distribution with 1 df.

numbers can be assigned before the third one is automatically set, there are 2 df for the chi-square test statistic in this case, and the distribution is shown in Figure 3.12. The statistic must be greater than 5.99 to cause rejection at the 5% significance level. The calculations in Table 3.9 show that the 5, 7, 8 split is far from causing rejection. It is interesting to note that the examples in both Tables 3.8 and 3.9 follow from the same set of 20 numbers:

22 (R)	13 (R)	32 (B)	31 (R)
16 (B)	9 (B)	25 (B)	26 (R)
33 (R)	8 (R)	28 (R)	11 (R)
17 (R)	29 (R)	20 (R)	22 (R)
2 (R)	4 (R)	29 (R)	13 (R)

Different conclusions are reached about "unbiased" by focusing on different measures of bias.

Although the chi-square goodness-of-fit test is easy to apply, it can give misleading results when expected counts are small. A category in which the expected count was 0.1 but the observed count was 1, for example, would contribute 8.1 to the test statistic and would be likely to lead to rejection of the hypothesis even though 1 is one of the two closest integers to 0.1. There have been several ad-hoc

Table 3.8 Goodness-of-fit calculations for two categories.

Category	Observed (o)	Expected (e)	$)o - e)$	$(o - e)^2/e$
Red	16	10	6	3.6
Black	4	10	-6	3.6
Total	20	20	0	7.2

Table 3.9 Goodness-of-fit calculations for three categories.

Category	Observed (o)	Expected (e)	$(o - e)$	$(o - e)^2/e$
première	5	6.67	-1.67	0.41
milieu	7	6.67	0.33	0.01
dernière	8	6.67	1.33	0.13
Total	20	20	0	0.55

rules put forward to reduce the chance of spurious significant results, but a better procedure is to avoid the chi-square goodness-of-fit test whenever small expected counts occur. The probability tests described below offer one means of avoidance.

Exercise 3.12 A roulette wheel gave 3 black numbers in 10 spins. Would you reject the hypothesis that the wheel was fair?

Exact Test

Now that computing power is widely available, many statistical tests are being conducted as *exact tests* or *probability tests* introduced by Fisher (1935). Briefly, these tests assume the hypothesis is true and calculate the probability of the observed outcome or a more extreme (less probable) outcome. Low values of this probability suggest that the hypothesis is not true.

Returning to the example of 16 reds in 20 spins of a roulette wheel, the probabilities of all 21 possible outcomes, grouped into ten pairs plus the most probable outcome, are shown in Table 3.10. We represent the number of reds by x, and because the binomial distribution is symmetrical in this case, we have

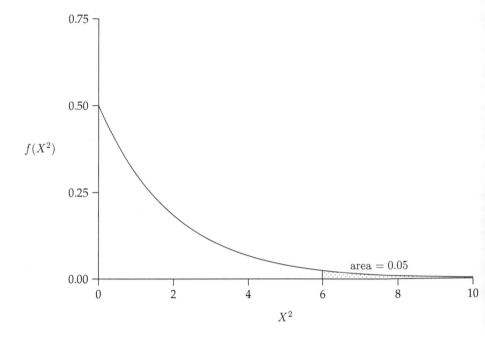

Figure 3.12   Chi-square distribution with 2 df.

chosen to group the 20 outcomes of $x \leq 9, x \geq 11$ into ten pairs: $(20,0)$, ... , $(11,9)$. The two outcomes in each pair have the same probability, so the probability of each pair is twice the probability of either member of the pair. The outcome of 16 reds has a probability of 0.0046 if the wheel is fair. This outcome, *or a more extreme outcome*, referring to the 10 outcomes of 20, 19, 18, 17, and 16 as well as 0, 1, 2, 3, and 4, has a 1.18% probability of occurring if the wheel is fair. Therefore the outcome of 16 belongs to the least probable 1.18% of the outcomes, and 0.0118 is called the P-value for the outcome.

The alternative to the hypothesis of fairness is two-tailed, meaning that we will reject if there are too few or two many reds. With this exact test, we are also using the two tails of the binomial distribution, unlike the use of just one tail of the chi-square distribution. A rejection region of 5% is constructed by looking for the most extreme 2.5% in each tail of the binomial. Both these tails of the binomial have outcomes a long way from the hypothesized value of x. For the chi-square distribution, only the upper tail has these extreme departures from the hypothesized value.

If we consider that 0.0118 is a low probability, then we would reject the hy-

It is the assumption of random mating, and the consequent independence, that allows the probability of each mate pair to be written in the third column as the product of the two separate genotype probabilities. The last three columns show, for each parental combination, the probability of each of the three possibilities for the children of that combination.

Now we introduce the term P'_{11} as the proportion of the $A_1 A_1$ genotypes among the children, i.e., the second generation. We can calculate P'_{11} from the terms in rows 1, 2, 4, and 5 of the table, using the law of total probability, as follows:

$$\begin{aligned} P'_{11} &= (1)P^2_{11} + (1/2)P_{11}P_{12} + (1/2)P_{12}P_{11} + (1/4)P^2_{12} \\ &= [P_{11} + (1/2)P_{12}]^2 \end{aligned}$$

Now we can apply Equation 4.1, which tells us that the expression in the brackets is simply p_1, so

$$P'_{11} = p^2_1$$

If we define P'_{12} as the proportion of $A_1 A_2$ children, then from rows 2 through 8 in Table 4.1

$$\begin{aligned} P'_{12} &= (1/2)P_{11}P_{12} + (1)P_{11}P_{22} + (1/2)P_{12}P_{11} + (1/2)P^2_{12} \\ &\quad + (1/2)P_{12}P_{22} + (1)P_{22}P_{11} + (1/2)P_{22}P_{12} \\ &= 2[P_{11} + (1/2)P_{12}][P_{22} + (1/2)P_{12}] \\ &= 2p_1 p_2 \end{aligned}$$

Similarly we can show that the proportion of $A_2 A_2$ children is $P'_{22} = p^2_2$. This shows that the offspring genotype proportions are specified completely by parental allele proportions as

$$P'_{11} = (p_1)^2, \quad P'_{12} = 2p_1 p_2, \quad P'_{22} = (p_2)^2 \tag{4.2}$$

This is a demonstration of the **Hardy-Weinberg law**. We can now use Equation 4.2 to calculate p'_1, the proportion of allele A_1 among the children:

$$\begin{aligned} p'_1 &= P'_{11} + \frac{1}{2}P'_{12} \\ &= p^2_1 + p_1 p_2 \\ &= p_1(p_1 + p_2) \\ &= p_1 \end{aligned}$$

The allele proportions are unchanged in the second generation. Note that the assumption of random mating has led to the Hardy-Weinberg law for the geno-types of the children, even though we made no assumption about relationships between the genotypic and allele proportions in the parental generation. We did invoke the counting rule of Equation 4.1, but that always holds for codominant alleles. For the more general case of an unspecified number of alleles the Hardy-Weinberg law is

$$\left.\begin{array}{rcl} P_{ii} & = & p_i^2 \\ P_{ij} & = & 2p_ip_j, \ \ j \neq i \end{array}\right\} \tag{4.3}$$

Note that we now have two separate ways of relating allelic and genotypic proportions: Equations 4.1 and 4.3. As we have seen, Equation 4.1 is always true and codominant allelic proportions can always be found from genotypic proportions in this way. Equation 4.3 enables genotypic proportions to be found as products of allelic proportions, but this should generally be regarded as only an approximation. The Hardy-Weinberg law was demonstrated above under the assumptions of random mating in an infinite population, without other forces such as selection, mutation, or migration. Under these circumstances, the law holds in all generations after the first, and so describes an *equilibrium* situation. As will be shown in the next section, however, the law may hold even if these circumstances do not occur, i.e., it can hold when there is selection (Lewontin and Cockerham 1959) or when there is nonrandom mating (Li 1988).

Exercise 4.2   Assuming Hardy-Weinberg equilibrium, find the three genotypic pro-portions for a gene with allele proportions of 0.7 and 0.3 for A_1 and A_2.

Exercise 4.3   Assume Hardy-Weinberg equilibrium to find the proportions of the four blood group types, A, B, AB, and O when the allele proportions are $p_A = 0.2, p_B = 0.1$, and $p_O = 0.7$.

DISTURBING FORCES

In the previous section we showed that, in infinite random-mating populations, allele proportions do not change from the parent to offspring generations and, if the conditions for Hardy-Weinberg equilibrium apply, then they will remain constant through all generations. Proportions can change if there are disturbing forces, such as selection, mutation, and migration.

Selection

Selection refers to the differential abilities of genotypes to contribute to the next generation. One mode of selection is *viability selection*, which involves the ability of an individual to survive to adulthood. If these abilities depend on the genotype of an individual, allele proportions will be altered. To illustrate the kinds of arguments that can be made, suppose the three genotypes A_1A_1, A_1A_2, and A_2A_2 at locus **A** have *viabilities* w_{11}, w_{12}, and w_{22}. This changes the genotype proportions from P_{11}, P_{12}, P_{22} at the beginning of the generation to P'_{11}, P'_{12}, and P'_{22} at the end of the generation, where

$$P'_{11} = w_{11}P_{11}/\bar{w}$$
$$P'_{12} = w_{12}P_{12}/\bar{w}$$
$$P'_{22} = w_{22}P_{22}/\bar{w}$$

Dividing by the *mean viability* $\bar{w}$ ensures that the proportions still add to one:

$$\bar{w} = w_{11}P_{11} + w_{12}P_{12} + w_{22}P_{22}$$

Allele proportions will change over time until an equilibrium is established. The equilibrium may reflect the loss of an unfavorable allele: p_1 will become zero if $w_{22} > w_{12} > w_{11}$ for example. In the special case of *heterozygote advantage*, $w_{11} < w_{12} > w_{22}$, it can be shown (Box 4.2) that a polymorphic equilibrium will be established. This is a situation where heterozygotes are the "fittest" and will always remain in the population. As long as they remain, it is guaranteed that both alleles will also remain in the population.

Lewontin and Cockerham (1959) showed that if $w_{11}w_{22} = w_{12}^2$ then the Hardy-Weinberg relation also holds after selection:

$$P'_{11} = (p'_1)^2, \ P'_{12} = 2p'_1p_2, \ P'_{22} = (p'_2)^2$$

Although this is a very restrictive situation, it makes the point that consistency with Hardy-Weinberg proportions does not mean that selection is absent.

In spite of this elegant theory, there are very few cases known where human genes exhibit these kinds of single-gene selective forces. The one case that recurs in textbooks is that of sickle-cell anemia. Sickle-cell hemoglobin is produced by a single substitution in the DNA sequence for β-hemoglobin. If A represents the normal form and S represents the sickle form, then AA individuals are normal. Affected homozygotes SS suffer from a severe hemolytic anemia, and their reproductive fitness is very low under primitive living conditions. The AS heterozygotes are clinically healthy, and it appears that they are at a reproductive advantage in the primitive conditions because they are protected against

---

Box 4.2: Allele proportion changes under selection

The allele proportions after selection are

$$
\begin{aligned}
p_1' &= P_{11}' + (P_{12}'/2) \\
&= [w_{11}P_{11} + (w_{12}P_{12}/2)]/\bar{w} \\
p_2' &= [w_{22}P_{22} + (w_{12}P_{12}/2)]/\bar{w}
\end{aligned}
$$

and if the original genotype proportions obey the Hardy-Weinberg law, $P_{11} = p_1^2, P_{12} = 2p_1p_2, P_{22} = p_2^2$,

$$
\begin{aligned}
p_1' &= p_1[w_{11}p_1 + w_{12}p_2]/\bar{w} \\
p_2' &= p_2[w_{22}p_2 + w_{12}p_2]/\bar{w}
\end{aligned}
$$

If an equilibrium is established, there is no change in proportion, $p_1' = p_1$, and the equilibrium value is written as $\hat{p}_1$. Then

$$
\bar{w} = w_{11}\hat{p}_1 + w_{12}\hat{p}_2 = w_{22}\hat{p}_2 + w_{12}\hat{p}_2
$$

and this can be rearranged to provide

$$
\left.
\begin{aligned}
\hat{p}_1 &= \frac{(w_{12} - w_{22})}{(w_{12} - w_{11}) + (w_{12} - w_{22})} \\[2ex]
\hat{p}_2 &= \frac{(w_{12} - w_{11})}{(w_{12} - w_{11}) + (w_{12} - w_{22})}
\end{aligned}
\right\}
\tag{4.4}
$$

These equilibrium allele proportions are valid, lying between zero and one, for heterozygote advantage: $(w_{12} - w_{11}) > 0, (w_{12} - w_{22}) > 0$.

Falciparum malaria. Some studies (Allison 1954) have estimated the following selection coefficients, measured relative to that for heterozygotes:

$$
\begin{aligned}
w_{AA} &= 0.7961 \\
w_{AS} &= 1.0000 \\
w_{SS} &= 0.1698
\end{aligned}
$$

Substituting these values into Equation 4.4 in Box 4.2 provides

$$
\hat{p}_S = \frac{0.2039}{0.8302 + 0.2039} = 0.1972
$$

This value has some support in empirical studies.

the same allele–i.e., they both descend from just one of the alleles received by H from his or her parents. We write the probability of this event of *identity by descent*, ibd, as

$$\Pr(h_1 \text{ is ibd to } h_2) = \Pr(h_1 \equiv h_2)$$
$$= 0.5$$

We will also use ibd for the phrase "identical by descent." Individuals X and Y, in turn, transmit alleles a and b to their child I, and we are interested in the probability of these two alleles being ibd. For this to occur, first h_1 and h_2 must be ibd, and then X must transmit a copy of h_1 (with probability 0.5), and Y a copy of h_2 (with probability 0.5):

$$\Pr(a \equiv b) = \Pr(a \equiv h_1, b \equiv h_2 | h_1 \equiv h_2) \Pr(h_1 \equiv h_2)$$
$$= \Pr(a \equiv h_1, b \equiv h_2) \Pr(h_1 \equiv h_2)$$
$$= \Pr(a \equiv h_1) \Pr(b \equiv h_2) \Pr(h_1 \equiv h_2)$$
$$= 0.5 \times 0.5 \times 0.5$$
$$= 0.125$$

Note that the two events of a being a copy of h_1 and b being a copy of h_2 are independent, and they do not depend on the event $h_1 \equiv h_2$. We define the probability that I receives a pair of ibd alleles from his or her parents as the *inbreeding coefficient* F_I of I. So, in this case,

$$F_I = \Pr(a \equiv b) = 0.125$$

Now if H had parents who were related, he or she would have a nonzero inbreeding coefficient F_H, and there would be two ways in which h_1 and h_2 could be ibd:

- With probability 0.5, they descended from the same single allele of the two alleles carried by H.

- With probability 0.5, they descended from the two different alleles in H, and with probability F_H, these alleles were ibd.

Therefore

$$\Pr(h_1 \equiv h_2) = \frac{1}{2} + \frac{1}{2} F_H$$

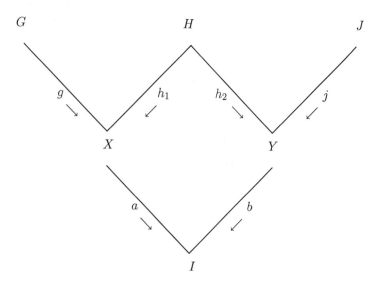

Figure 4.2 I is child of half sibs X and Y

so that, following the same argument as before,

$$
\begin{aligned}
\Pr(a \equiv b) &= \Pr(a \equiv h_1, b \equiv h_2 | h_1 \equiv h_2) \Pr(h_1 \equiv h_2) \\
&= \Pr(a \equiv h_1, b \equiv h_2) \Pr(h_1 \equiv h_2) \\
&= \Pr(a \equiv h_1) \Pr(b \equiv h_2) \Pr(h_1 \equiv h_2) \\
&= 0.5 \times 0.5 \times (1 + F_H)/2 \\
&= (1 + F_H)/8
\end{aligned}
$$

A general approach is to specify some initial or reference population, in which all members are assumed to be unrelated, and then to measure inbreeding relative to that generation. It is generally accepted, for example, that Finland was settled by a relatively small group of people about 4,000 years ago. It would be convenient to quantify inbreeding for the present population as the probability that a random person in the population (assumed to have descended from the initial group) receives two alleles that trace back to a single allele among the founders. Alleles that trace to distinct founding alleles will be considered not ibd since we assumed there was no relatedness among the founders.

The argument given for half sibs leads to *path-counting* equations for inbreeding coefficients. Suppose the parents X and Y of individual I have a common ancestor A. One of these ancestors is shown in Figure 4.3, although there may be several and they need not all be in the same generation. Also suppose that

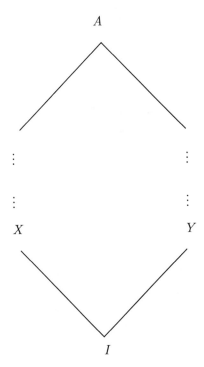

A

X   Y

I

Figure 4.3   The parents X and Y of I have a common ancestor A.

there are n_A people in the loop from one parent through A and back to the other parent. Then, summing over all common ancestors A of X and Y, the inbreeding coefficient of I is

$$F_I = \sum_A \left(\frac{1}{2}\right)^{n_A} (1 + F_A) \tag{4.6}$$

Each person in the loop, apart from A but including the parents X and Y, introduces another probability of 0.5 for the transmission of an allele from A. The two alleles that common ancestor A gives to the two sides of the loop have a probability $(1 + F_A)/2$ of being ibd.

In the half sib case of Figure 4.2, parent H is the only common ancestor of X and Y and $n_H = 3$ for the path XHY, so $F_I = 1/8$ as before. In Figure 4.4, full sibs X and Y have two parents G and H in common. The two paths are XGY and XHY, each with three individuals, so the inbreeding coefficient of their child I would be $(1/2)^3 + (1/2)^3 = 1/4$, providing G and H are not inbred. In Figure 4.5, first cousins X and Y have four distinct parents, G, H, J and K, two of whom are full sibs, so they have two grandparents A and B in common. The two paths $XHAJY$ and $XHBJY$ each have five individuals. The inbreeding coefficient of the children of first cousins is therefore $(1/2)^5 + (1/2)^5 = 1/16$, and this is the maximum amount of inbreeding tolerated by most marriage laws.

Just as the concepts of inbreeding and relatedness are closely connected, so are the probabilities of these events. The usual measure of relatedness for individuals X and Y is the *coancestry coefficient* θ_{XY}, defined as the probability that two alleles, one taken at random from each of X and Y, are ibd. If a and b are the alleles from X and Y, then

$$\theta_{XY} = \Pr(a \equiv b | a, b \text{ have come from } X \text{ and } Y)$$

If individuals X and Y have a child I,

$$F_I = \theta_{XY} \tag{4.7}$$

so we have shown in the preceding examples that θ_{XY} is $1/4$ for full sibs, $1/8$ for half sibs, and $1/16$ for first cousins.

There is one additional relation needed to characterize inbreeding in pedigrees or in populations. That is the probability of two alleles from the same individual being ibd. For individual X we have already defined F_X as the probability that X *receives* two ibd alleles. Now we introduce the probability θ_{XX} that X *transmits* two ibd alleles, and this is

$$\theta_{XX} = \Pr(a \equiv b | a, b \text{ are both from } X)$$

In the examples we have done so far, we have assumed that each individual is not inbred, and if X is not inbred then $\theta_{XX} = 0.5$. However, there will be occasions where we need to allow for inbred individuals. Consider individual X in Figure 4.6, who receives alleles c and d from his parents and who transmits alleles a and b to two of his children. There are four possibilities, each of which has probability $1/4$:

- a and b are both copies of c.

- a is a copy of c, and b is a copy of d.

- a is a copy of d, and b is a copy of c.

- a and b are both copies of d.

Using the law of total probability,

$$\theta_{XX} = \Pr(a \equiv b)$$
$$= \frac{1}{4}[\Pr(c \equiv c) + \Pr(c \equiv d) + \Pr(d \equiv c) + \Pr(d \equiv d)]$$

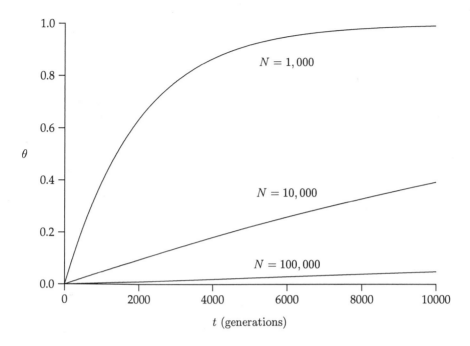

Figure 4.7   Change in θ over time with drift.

Exercise 4.7   Calculate the actual inbreeding coefficient for the simulations in Exercise 4.6 as the proportion of individuals with ibd alleles in each generation. To do this, take alleles 1 and 2 to represent individual 1, alleles 3 and 4 to represent individual 2, and so on. The theoretical values are 0.00, 0.10, 0.19, 0.27, 0.34, and 0.41 for $t = 0, 1, 2, 3, 4$, and 5.

The calculations in Box 4.7 show that the Equation 4.9 is approximately true even when selfing is not possible. Numerical results for the two sets of expressions are shown in Table 4.3. It can be seen that, for large population sizes, there is no discernible difference in the inbreeding levels for the two mating systems. More elaborate theory can be developed for separate sexes, but the result is the same - population genetic theory for human populations can be approximated by theory for monoecious populations.

Genetic drift, unlike mutation or selection, does not change the overall average, or **expected**, allele proportions. The average proportion of an allele over many replicate populations (see Figure 4.1) remains the same, although it will certainly change within any single population. There is therefore a great deal of

Box 4.7: Avoidance of selfing

What relevance does Equation 4.9 have for human populations, where selfing is not possible? As a partial demonstration that it is a very good approximation, consider a population of size N where mating is at random but selfing is not possible. A child in generation $t + 1$ must receive alleles from distinct parents in generation t, and these alleles have chance θ of being ibd:

$$F_{t+1} = \theta_t$$

Alleles received by different children, however, can come from the same or different parents with probabilities $1/N$ and $(N - 1)/N$, so that

$$\theta_{t+1} = \frac{1}{N}\frac{1 + F_t}{2} + \frac{N - 1}{N}\theta_t$$

Putting these two equations together provides

$$F_{t+2} = \frac{1}{2N} + \frac{N - 1}{N}F_{t+1} + \frac{1}{2N}F_t$$

For the case where individuals in the initial population are noninbred and unrelated, this leads to

$$F_t = 1 - \left[\frac{(1 - \lambda_2)\lambda_1^t - (1 - \lambda_1)\lambda_2^t}{\lambda_1 - \lambda_2}\right]$$

Here $\lambda_1, \lambda_2 = [(N - 1)/N \pm \sqrt{(1 + 1/N^2)}]/2$ and when N is large, $\lambda_1 \approx (1 - 1/2N)$, $\lambda_2 \approx 0$. In that case the result reduces to Equation 4.9, derived for random mating when selfing is allowed.

Table 4.3   Inbreeding coefficients for monoecious populations of size $N = 100,000$.

Generation	With selfing	Without selfing	
t	$F = \theta$	F	θ
1,000	0.0050	0.0050	0.0050
10,000	0.0488	0.0488	0.0488
100,000	0.3935	0.3935	0.3935
1,000,000	0.9933	0.9933	0.9933
10,000,000	1.0000	1.0000	1.0000

variation generated between populations. A series of finite isolated populations, each descending from the same reference population, will drift apart over time.

Genotype Proportions in Inbred Populations

Another way of expressing the phenomenon of genetic drift is to express genotypic proportions in terms of allelic proportions *and* the inbreeding coefficient. An individual will be homozygous if it receives two copies of the same allele. With probability F these two are ibd, and therefore are both allele A if either one is A. With probability $(1 - F)$ the two are not ibd, and each has chance p_A of being allele A:

$$P_{AA} = Fp_A + (1 - F)p_A^2$$

or, generally,

$$\left. \begin{aligned} P_{ii} &= p_i^2 + p_i(1 - p_i)F \\ P_{ij} &= 2p_ip_j(1 - F) \end{aligned} \right\} \tag{4.10}$$

showing that genotypic proportions can be expressed as the Hardy-Weinberg values plus a deviation due to drift. In fact, this deviation can be due to any system of *inbreeding*–meaning a mating system in which an individual can receive ibd allele pairs. Note that these equations assume the initial population to be in Hardy-Weinberg equilibrium. For random-mating populations, the ibd status of pairs of alleles is the same whether they are carried by the same or different individuals, and then $F = \theta$. For this reason, Equations 4.10 are sometimes written with θ replacing F.

For alleles that are both harmful and recessive, such as the $\Delta F508$ allele responsible for most cases of cystic fibrosis, inbreeding increases the proportion of people with the harmful trait by virtue of having two copies of the deleterious allele. These two alleles are not masked by a normal allele. The $\Delta F508$ allele in Caucasian populations has a proportion of about $p = 0.05$. Among individuals whose parents are unrelated, the probability of having two copies of the allele, and therefore having cystic fibrosis, is about $p^2 = 0.0025$. Among people whose parents are cousins, however, with probability $(1 - F) = 15/16$ the genotype probability is p^2, and with probability $F = 1/16$ it has the higher value of p. The total probability of the disease among these inbred people is more than doubled, to 0.0055.

Homozygotes have two alleles that have the same chemical composition, and so are *identical in state*. Such alleles may or may not be ibd. Heterozygotes have alleles that are not identical in state, and these alleles cannot be ibd.

Drift and Mutation

A previous section showed that mutation and selection could act in opposite ways to lead to an equilibrium proportion for an allele. A different type of equilibrium can be established between drift and mutation. Genetic variation is lost by drift, as alleles tend to become *fixed*, whereas variation can continually be introduced by mutation of the form that every mutant is a new type of allele. Equilibrium now refers to a constant amount of variation–the proportion of any particular allele will be changing. Every allele introduced by mutation will eventually be lost or fixed. Any fixation is temporary, however, as further mutations will introduce other alleles.

A convenient way to characterize such populations is by use of the inbreeding coefficient. The transition equation for changes due to drift

$$F' \;=\; \frac{1}{2N} + (1 - \frac{1}{2N})F$$

has to be modified. Alleles can remain ibd only if neither of them mutates, and this happens with probability $(1 - \mu)^2$, where μ is the mutation rate. Therefore

$$F' \;=\; (1 - \mu)^2 \left[\frac{1}{2N} + (1 - \frac{1}{2N})F \right]$$

At equilibrium, the inbreeding coefficient is no longer changing and has the value

$$\hat{F} \;=\; \frac{(1 - \mu)^2/2N}{1 - (1 - \mu)^2(1 - 1/2N)}$$

$$\approx \frac{1}{1 + 4N\mu}$$

An alternative expression is

$$\hat{H} \;\approx\; \frac{4N\mu}{1 + 4N\mu}$$

where $\hat{H}$ is the proportion of heterozygotes in the equilibrium population.

FOUR-ALLELE DESCENT MEASURES

So far, we have considered the genotypes of individuals or of collections of individuals. There are occasions, such as in the "brother's defense" discussed later, where it is necessary to consider the genotypes of pairs of individuals. These pairs may be relatives, such as brothers, or they may both belong to some specific subpopulation. For pairs of people, it is necessary to introduce more ibd

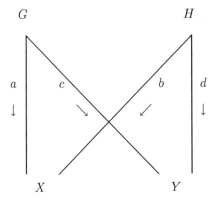

Figure 4.8   Pedigree of full sibs X and Y.

measures. Whereas the genotypic proportion of possibly inbred individuals requires the use of a descent measure defined for pairs of alleles, the genotypic
proportions for pairs of relatives require descent measures for three or four alleles being ibd. Suppose individual X has alleles a and b at locus **A** and Y has
alleles c and d. There are fifteen possible ibd relations among the four alleles, as
shown in Table 4.4, along with their probabilities δ. The fifteen probabilities add
to one.

We need to explain the notation. For each δ, the subscript indicates which
alleles are ibd, and alleles not in the subscript are neither ibd to each other nor to
the alleles specified. For example, the quantity δ_{ab} is the probability that $a \equiv b$ are
ibd *and* that c and d are neither ibd to each other nor to a and b. The equivalence
sign $\equiv$ is being used to indicate ibd. The quantity $\delta_{ab.cd}$ is the probability that
$a \equiv b, c \equiv d$ but that the two pairs a, b and c, d are not ibd to each other. Finally,
δ_{abcd} is the probability that all four alleles are ibd.

As an example consider the case where X and Y are full sibs with parents G
and H, as was shown in Figure 4.4. We redraw the pedigree in Figure 4.8 to show
the four alleles a, b, c and d. If G and H are not inbred and are unrelated, there
are only four possible ibd relationships with nonzero probabilities: $a \equiv c, b \equiv d$;
$a \equiv c, b \not\equiv d; a \not\equiv c, b \equiv d$; and $a \not\equiv c, b \not\equiv d$. Each of these has probability $1/4$,
as shown in Table 4.4.

What is the relationship between the four-allele descent measures δ and the
two-allele measure θ introduced earlier? Recall that θ_{XY} was defined as the
probability of a random allele from X being ibd to a random allele from Y.

Table 4.4   Descent relations among alleles for two individuals: X with alleles a and b and Y with alleles c and d.

Alleles ibd[1]	Probability	
	General	Full sibs[2]
none	δ_o	1/4
$a \equiv b$	δ_{ab}	0
$c \equiv d$	δ_{cd}	0
$a \equiv c$	δ_{ac}	1/4
$a \equiv d$	δ_{ad}	0
$b \equiv c$	δ_{bc}	0
$b \equiv d$	δ_{bd}	1/4
$a \equiv b \equiv c$	δ_{abc}	0
$a \equiv b \equiv d$	δ_{abd}	0
$a \equiv c \equiv d$	δ_{acd}	0
$b \equiv c \equiv d$	δ_{bcd}	0
$a \equiv b, c \equiv d$	$\delta_{ab.cd}$	0
$a \equiv c, b \equiv d$	$\delta_{ac.bd}$	1/4
$a \equiv d, b \equiv c$	$\delta_{ad.bc}$	0
$a \equiv b \equiv c \equiv d$	δ_{abcd}	0
Total	1	1

[1] Alleles not specified are not ibd.
[2] a, c from mother; b, d from father.

When X has alleles a and b and Y has alleles c and d this means that

$$\theta_{XY} = \frac{1}{4}[\Pr(a \equiv c) + \Pr(a \equiv d) + \Pr(b \equiv c) + \Pr(b \equiv d)]$$

so that

$$\begin{aligned}
\theta_{XY} = \frac{1}{4}[&(\delta_{ac} + \delta_{abc} + \delta_{acd} + \delta_{ac.bd} + \delta_{abcd}) \\
&+ (\delta_{ad} + \delta_{abd} + \delta_{acd} + \delta_{ad.bc} + \delta_{abcd}) \\
&+ (\delta_{bc} + \delta_{abc} + \delta_{bcd} + \delta_{bc.ad} + \delta_{abcd}) \\
&+ (\delta_{bd} + \delta_{abd} + \delta_{bcd} + \delta_{bd.ac} + \delta_{abcd})]
\end{aligned} \qquad (4.11)$$

Noninbred Relatives

Now suppose that neither of two individuals $X(a, b)$ and $Y(c, d)$ is inbred. Then any probability involving $a \equiv b$ or $c \equiv d$ is zero, and there are only seven measures to consider: $\delta_0, \delta_{ac}, \delta_{ad}, \delta_{bc}, \delta_{bd}, \delta_{ac.bd}$, and $\delta_{ad.bc}$. Furthermore, if there is symmetry between a, b and c, d then there are only three distinct values for these seven measures, according to whether the two individuals have zero (δ_0), or one ($\delta_{ac}, \delta_{ad}, \delta_{bc}, \delta_{bd}$), or two ($\delta_{ac.bd}, \delta_{ad.bc}$) alleles identical by descent. There are occasions when it is better to work with all seven δ measures, however.

For noninbred relatives, Equation 4.11 reduces to

$$\theta_{XY} \;=\; \frac{1}{4}[(\delta_{ac} + \delta_{ad} + \delta_{bc} + \delta_{bd}) + 2(\delta_{ac.bd} + \delta_{ad.bc})]$$

For full sibs, therefore (as before),

$$\theta_{XY} \;=\; \frac{1}{4}[(\frac{1}{4} + 0 + 0 + \frac{1}{4}) + 2(\frac{1}{4} + 0)] = \frac{1}{4}$$

Another useful summary measure in the noninbred case is the two-allele-pair measure $\Delta_{\ddot{X}+\ddot{Y}}$. This is the average of the two probabilities that X and Y have two pairs of ibd alleles:

$$\Delta_{\ddot{X}+\ddot{Y}} \;=\; \frac{1}{2}(\delta_{ac.bd} + \delta_{ad.bc})$$

When X and Y are full sibs,

$$\Delta_{\ddot{X}+\ddot{Y}} \;=\; \frac{1}{2}(\frac{1}{4} + 0) = \frac{1}{8}$$

The probabilities that noninbred relatives X and Y have 0, 1, or 2 pairs of ibd alleles can be summarized as

$$\left.\begin{array}{ll} \text{0 pairs}: & 1 - 4\theta_{XY} + 2\Delta_{\ddot{X}+\ddot{Y}} \\ \text{1 pair}: & 4(\theta_{XY} - \Delta_{\ddot{X}+\ddot{Y}}) \\ \text{2 pairs}: & 2\Delta_{\ddot{X}+\ddot{Y}} \end{array}\right\} \qquad (4.12)$$

Joint Genotypic Probabilities

The descent status of the four alleles that two individuals have between them puts constraints on the possible genotypes of the individuals (Cockerham 1971). If all four alleles were ibd, for example, both individuals would need to be of the same homozygous genotype. The converse relation is more complicated because homozygous individuals need not have alleles that are ibd. We now consider all possible pairs of genotypes for individuals related to any degree (specified by the descent measures in Table 4.4)

Two homozygotes.   What is the probability that X and Y are both homozygous $A_i A_i$? In the third column of Table 4.5 we show contributions to this probability from each of the 15 ibd relationships. With probability δ_0, there is no identity by descent among the four alleles, so each is of independent origin, and each has probability p_i of being of type A_i. In the second row of Table 4.5, only alleles a and b are ibd and have the same origin. This means that there are three alleles with independent origin: ab, c, and d, and each has probability p_i of being of type A_i. The probability of two $A_i A_i$ genotypes in this ibd situation is therefore p_i^3. The overall probability is arrived at by combining third-column terms over rows of Table 4.5 using the law of total probability, and is

$$\Pr(A_i A_i, A_i A_i) \;=\; \delta_{abcd}p_i + (\delta_{abc} + \delta_{abd} + \delta_{acd} + \delta_{bcd})p_i^2$$
$$+ (\delta_{ab.cd} + \delta_{ac.bd} + \delta_{ad.bc})p_i^2$$
$$+ (\delta_{ab} + \delta_{ac} + \delta_{ad} + \delta_{bc} + \delta_{bd} + \delta_{cd})p_i^3 + \delta_0 p_i^4$$

The probability with which two individuals are homozygous, but for different alleles, can be derived in a similar manner, using the fourth column of Table 4.5. For any of the rows in which the individuals share ibd alleles, there is zero probability that they could be homozygous for different alleles. Adding the fourth-column terms over rows gives

$$\Pr(A_i A_i, A_j A_j) \;=\; \delta_{ab.cd}p_i p_j + \delta_{ab}p_i p_j^2 + \delta_{cd}p_i^2 p_j + \delta_0 p_i^2 p_j^2$$

When neither X nor Y is inbred, the last two results simplify to

$$\Pr(A_i A_i, A_i A_i) \;=\; p_i^4 + 4\theta_{XY}p_i^3(1 - p_i) + 2\Delta_{\ddot{X}+\ddot{Y}}p_i^2(1 - p_i)^2 \qquad (4.13)$$

$$\Pr(A_i A_i, A_j A_j) \;=\; (1 - 4\theta_{XY} + 2\Delta_{\ddot{X}+\ddot{Y}})p_i^2 p_j^2 \qquad (4.14)$$

Under the assumption of no inbreeding, it is quicker to derive these last two results directly from Equations 4.12.

One homozygote and one heterozygote.   For one homozygous and one heterozygous individual, we need to distinguish between the cases when the individuals share an allele and when they do not. We also need to allow for both orderings of alleles within heterozygotes, and the calculations follow from Table 4.6.

Adding columns 2 and 3 in Table 4.6, multiplying by column 1, and then summing over rows, we get

$$\Pr(A_i A_i, A_i A_j) \;=\; 2\delta_0 p_i^3 p_j + (2\delta_{ab} + \delta_{ac} + \delta_{ad} + \delta_{bc} + \delta_{bd})p_i^2 p_j$$
$$+ (\delta_{abc} + \delta_{abd})p_i p_j$$

Table 4.5 Probabilities of $X(a,b)$ and $Y(c,d)$ being the same homozygote A_iA_i or different homozygotes A_iA_i, A_jA_j.

IBD[1]	Pr(IBD)	Pr(A_iA_i, A_iA_i|IBD)	Pr(A_iA_i, A_jA_j|IBD))
none	δ_o	p_i^4	$p_i^2 p_j^2$
$a \equiv b$	δ_{ab}	p_i^3	$p_i p_j^2$
$c \equiv d$	δ_{cd}	p_i^3	$p_i^2 p_j$
$a \equiv c$	δ_{ac}	p_i^3	0
$a \equiv d$	δ_{ad}	p_i^3	0
$b \equiv c$	δ_{bc}	p_i^3	0
$b \equiv d$	δ_{bd}	p_i^3	0
$a \equiv b \equiv c$	δ_{abc}	p_i^2	0
$a \equiv b \equiv d$	δ_{abd}	p_i^2	0
$a \equiv c \equiv d$	δ_{acd}	p_i^2	0
$b \equiv c \equiv d$	δ_{bcd}	p_i^2	0
$a \equiv b, c \equiv d$	$\delta_{ab.cd}$	p_i^2	$p_i p_j$
$a \equiv c, b \equiv d$	$\delta_{ac.bd}$	p_i^2	0
$a \equiv d, b \equiv c$	$\delta_{ad.bc}$	p_i^2	0
$a \equiv b \equiv c \equiv d$	δ_{abcd}	p_i	0

[1]IBD : ibd relationship

and adding over rows for column 1 multiplied by column 4 (doubled to take account of both heterozygote orders A_jA_k, A_kA_j), we get

$$\Pr(A_iA_i, A_jA_k) = 2\delta_0 p_i^2 p_j p_k + \delta_{ab} p_i p_j p_k$$

If X is not inbred, and since heterozygote Y cannot be inbred, these two equations become

$$\Pr(A_iA_i, A_iA_j) = 2(1 - 4\theta_{XY} + 2\Delta_{\ddot{X}+\ddot{Y}})p_i^3 p_j$$
$$+ 4(\theta_{XY} - \Delta_{\ddot{X}+\ddot{Y}})p_i^2 p_j \tag{4.15}$$

$$\Pr(A_iA_i, A_jA_k) = 2(1 - 4\theta_{XY} + 2\Delta_{\ddot{X}+\ddot{Y}})p_i^2 p_j p_k \tag{4.16}$$

and these results also follow directly from Equations 4.12.

The same two heterozygotes. When two individuals have the same heterozygous genotype, our approach requires an accounting for all four orders of the two alleles within each. Combining the last four columns of Table 4.7, multiplying by

Table 4.6   Probabilities of $X(a,b)$ being homozygous $A_i A_i$ and $Y(c,d)$ being heterozygous $A_i A_j$ or $A_j A_k$.

Pr(IBD)	Pr($A_i A_i, A_i A_j$|IBD)		Pr($A_i A_i, A_j A_k$|IBD))
	$c = A_i, d = A_j$	$c = A_j, d = A_i$	
δ_0	$p_i^3 p_j$	$p_i^3 p_j$	$p_i^2 p_j p_k$
δ_{ab}	$p_i^2 p_j$	$p_i^2 p_j$	$p_i p_j p_k$
δ_{cd}	0	0	0
δ_{ac}	$p_i^2 p_j$	0	0
δ_{ad}	0	$p_i^2 p_j$	0
δ_{bc}	$p_i^2 p_j$	0	0
δ_{bd}	0	$p_i^2 p_j$	0
δ_{abc}	$p_i p_j$	0	0
δ_{abd}	0	$p_i p_j$	0
δ_{acd}	0	0	0
δ_{bcd}	0	0	0
$\delta_{ab.cd}$	0	0	0
$\delta_{ac.bd}$	0	0	0
$\delta_{ad.bc}$	0	0	0
δ_{abcd}	0	0	0

column 1, and summing over rows leads to

$$
\begin{aligned}
\Pr(A_i A_j, A_i A_j) = {} & 4\delta_0 p_i^2 p_j^2 \\
& + (\delta_{ac} + \delta_{ad} + \delta_{bc} + \delta_{bd}) p_i p_j (p_i + p_j) \\
& + 2(\delta_{ac.bd} + \delta_{ad.bc}) p_i p_j
\end{aligned}
$$

This can be rewritten in a way that reflects the fact that heterozygotes cannot be inbred:

$$
\begin{aligned}
\Pr(A_i A_j, A_i A_j) = {} & 4(1 - 4\theta_{XY} + 2\Delta_{\ddot{X}+\ddot{Y}}) p_i^2 p_j^2 \\
& + 4(\theta_{XY} - \Delta_{\ddot{X}+\ddot{Y}}) p_i p_j (p_i + p_j) \\
& + 4\Delta_{\ddot{X}+\ddot{Y}} p_i p_j
\end{aligned}
\tag{4.17}
$$

This equation could also be derived directly from Equations 4.12.

Two different heterozygotes.   Finally we consider the case of two heterozygotes that have one or no alleles in common. With one allele in common,

$$
\begin{aligned}
\Pr(A_i A_j, A_i A_k) = {} & 4\delta_0 p_i^2 p_j p_k \\
& + (\delta_{ac} + \delta_{ad} + \delta_{bc} + \delta_{bd}) p_i p_j p_k
\end{aligned}
$$

Table 4.7   Probabilities of $X(a,b)$ and $Y(c,d)$ being heterozygous A_iA_j.

Pr(IBD)	$\Pr(A_iA_j, A_iA_j\vert\text{IBD})$			
	$a = A_i, b = A_j$ $c = A_i, d = A_j$	$a = A_i, b = A_j$ $c = A_j, d = A_i$	$a = A_j, b = A_i$ $c = A_i, d = A_j$	$a = A_j, b = A_i$ $c = A_j, d = A_i$
δ_0	$p_i^2p_j^2$	$p_i^2p_j^2$	$p_i^2p_j^2$	$p_i^2p_j^2$
δ_{ab}	0	0	0	0
δ_{cd}	0	0	0	0
δ_{ac}	$p_ip_j^2$	0	0	$p_i^2p_j$
δ_{ad}	0	$p_ip_j^2$	$p_i^2p_j$	0
δ_{bc}	0	$p_i^2p_j$	$p_ip_j^2$	0
δ_{bd}	$p_i^2p_j$	0	0	$p_ip_j^2$
δ_{abc}	0	0	0	0
δ_{abd}	0	0	0	0
δ_{acd}	0	0	0	0
δ_{bcd}	0	0	0	0
$\delta_{ab.cd}$	0	0	0	0
$\delta_{ac.bd}$	p_ip_j	0	0	p_ip_j
$\delta_{ad.bc}$	0	p_ip_j	p_ip_j	0
δ_{abcd}	0	0	0	0

which, after recognizing that both individuals must be noninbred, gives

$$\Pr(A_iA_j, A_iA_k) = 4(1 - 4\theta_{XY} + 2\Delta_{\ddot{X}+\ddot{Y}})p_i^2p_jp_k$$
$$+ 4(\theta_{XY} - \Delta_{\ddot{X}+\ddot{Y}})p_ip_jp_k \qquad (4.18)$$

When the two heterozygotes share no alleles, only the complete nonidentity probability δ_0 is needed. Multiplying by four to account for all four orders of alleles within individuals gives us

$$\Pr(A_iA_j, A_kA_l) = 4\delta_0p_ip_jp_kp_l$$

and, because the individuals are both noninbred,

$$\Pr(A_iA_j, A_kA_l) = 4(1 - 4\theta_{XY} + 2\Delta_{\ddot{X}+\ddot{Y}})p_ip_jp_kp_l$$

as may also be seen from Equations 4.12.

Genotypes of Two Sibs

As an application of these results, consider the possible genotype pairs for two sibs X and Y that have noninbred and unrelated parents, so they themselves

LIVERPOOL JOHN MOORES UNIVERSITY
LEARNING SERVICES

are not inbred. The descent measure values are shown in Table 4.4, and the probabilities for any of the six possible pairs of genotypes (regardless of order) can be found from Equations 4.13 to 4.18:

$$
\begin{array}{ll}
A_i A_i, A_i A_i & p_i^2 (1 + p_i)^2 / 4 \\
A_i A_i, A_j A_j & p_i^2 p_j^2 / 2 \\
A_i A_i, A_i A_j & p_i^2 p_j (1 + p_i) \\
A_i A_i, A_j A_k & 2 p_i^2 p_j p_k \\
A_i A_j, A_i A_j & p_i p_j (1 + p_i + p_j + 2 p_i p_j) / 2 \\
A_i A_j, A_i A_k & p_i p_j p_k (1 + 2 p_i) \\
A_i A_j, A_k A_l & 2 p_i p_j p_k p_l
\end{array}
$$

These are not the same as simply multiplying together the proportions of each of the two genotypes. Relatives are more likely than random members of the population to have the same genotype. Adding together the first and fifth of these equations and then summing over all alleles A_i, A_j gives the probability that two (untyped) sibs have the same genotype without specifying that genotype. This probability is

$$
\Pr(\text{sibs the same}) = \frac{1}{4} \left(1 + 2 \sum_i p_i^2 + 2 \left(\sum_i p_i^2 \right)^2 - \sum_i p_i^4 \right)
$$

This expression tends to $1/4$ as the number of alleles increases and the proportion of each one becomes small.

Exercise 4.8   Find the probability with which a noninbred uncle and his noninbred nephew both have $A_1 A_2$ genotypes if they belong to a population in Hardy-Weinberg equilibrium with proportions p_1 and p_2 for alleles A_1 and A_2.

MATCH PROBABILITIES

We now return to an assumption that we made when considering transfer evidence in Chapter 2. Recall that we arrived at Equation 2.3 as the expression for the LR in a single crime scene stain case:

$$
\text{LR} = \frac{1}{\Pr(G_C | G_S, H_d, I)}
$$

We assumed that knowledge of the suspect's genotype G_S did not influence our uncertainty about the offender's genotype. Cases arise, however, where that assumption is not correct. It is not correct, for example, when it is suggested that the offender is a close relative of the suspect. It is also not correct when the suspect and offender should realistically be considered both members of the same

and this is one-fourth the probability with which two individuals, X and Y, are heterozygous $A_i A_j$ (allowing for the two possible orders of alleles per individual). Combining Equations 4.24 and 4.26 gives

$$\Pr(G_X = A_i A_i | G_Y = A_i A_i) = \frac{\Pr(A_i^4)}{\Pr(A_i^2)}$$

$$= \frac{[(1 - \theta)p_i + 2\theta][(1 - \theta)p_i + 3\theta]}{(1 + \theta)(1 + 2\theta)}$$

as has been given in Equation 4.20. The conditional probability for heterozygotes $\Pr(G_X = A_i A_j | G_Y = A_i A_j)$, also given in Equation 4.20, follows from dividing Equation 4.27 by Equation 4.25.

The full power of the Dirichlet approach becomes apparent for sets of more than four alleles, for which there is no exact descent measure formulation. To anticipate a result needed in Chapter 6, the probability of a woman and an alleged father, if he is not the child's father, both being homozygous $A_i A_i$ and the woman's child receiving paternal allele A_i is

$$\Pr(A_i A_i, A_i A_i, A_i) = \Pr(A_i^5)$$

$$= \frac{\gamma_i(\gamma_i + 1)(\gamma_i + 2)(\gamma_i + 3)(\gamma_i + 4)}{\gamma.(\gamma. + 1)(\gamma. + 2)(\gamma. + 3)(\gamma. + 4)}$$

$$= \frac{p_i[(1 - \theta)p_i + \theta][(1 - \theta)p_i + 2\theta]}{(1 - \theta)(1 + \theta)}$$

$$\times \frac{[(1 - \theta)p_i + 3\theta][(1 - \theta)p_i + 4\theta]}{(1 + 2\theta)(1 + 3\theta)}$$

A result of the type needed in Chapter 7 is for the interpretation of a mixed stain having alleles A_i, A_j, and A_k from a victim and her attacker. If the victim is heterozygous $A_i A_j$ and a man suspected of being the attacker is heterozygous $A_i A_k$, a probability needed if the actual attacker is homozygous $A_k A_k$ is

$$\Pr(A_i A_j, A_i A_k, A_k A_k) = \Pr(A_i^2 A_j A_k^3)$$

$$= \frac{\gamma_i(\gamma_i + 1)\gamma_j\gamma_k(\gamma_k + 1)(\gamma_k + 2)}{\gamma.(\gamma. + 1)(\gamma. + 2)(\gamma. + 3)(\gamma. + 4)(\gamma. + 5)}$$

$$= \frac{p_i[(1 - \theta)p_i + \theta]p_j}{(1 + \theta)}$$

$$\times \frac{p_k[(1 - \theta)p_k + \theta][(1 - \theta)p_k + 2\theta)]}{(1 + 2\theta)(1 + 3\theta)(1 + 4\theta)}$$

Table 4.9   Gamete probabilities from genotypes at two loci.

Genotype	Gametic arrangement	Gamete A_1B_1	A_1B_2	A_2B_1	A_2B_2
$A_1A_1B_1B_1$	A_1B_1/A_1B_1	1	0	0	0
$A_1A_1B_1B_2$	A_1B_1/A_1B_2	1/2	1/2	0	0
$A_1A_1B_2B_2$	A_1B_2/A_1B_2	0	1	0	0
$A_1A_2B_1B_1$	A_1B_1/A_2B_1	1/2	0	1/2	0
$A_1A_2B_1B_2$	A_1B_1/A_2B_2	$(1-c)/2$	$c/2$	$c/2$	$(1-c)/2$
$A_1A_2B_1B_2$	A_1B_2/A_2B_1	$c/2$	$(1-c)/2$	$(1-c)/2$	$c/2$
$A_1A_2B_2B_2$	A_1B_2/A_2B_2	0	1/2	0	1/2
$A_2A_2B_1B_1$	A_2B_1/A_2B_1	0	0	1	0
$A_2A_2B_1B_2$	A_2B_1/A_2B_2	0	0	1/2	1/2
$A_2A_2B_2B_2$	A_2B_2/A_2B_2	0	0	0	1

PAIRS OF LOCI

All the theory to this point has been for a single genetic locus, whereas the human genome contains many thousands of loci. For purposes of human identification, it will obviously be better to use as much information, meaning as many loci, as possible. Some of the complications in going from one to many loci can be illustrated by considering two loci.

Consider locus **A** with alleles A_1 and A_2 and locus **B** with alleles B_1, and B_2. At locus **A** an individual may have genotype A_1A_1, A_1A_2, or A_2A_2, and at locus **B** the same individual may have genotype B_1B_1, B_1B_2, or B_2B_2. Taking both loci into account, there will therefore be nine different genotypes: $A_1A_1B_1B_1$, $A_1A_1B_1B_2$, ... , $A_1A_2B_2B_2$. These are displayed in Table 4.9. Individuals transmit one allele per locus to their children, so that there are four possible types of *gamete* or *haplotype*: A_1B_1, A_1B_2, A_2B_1, and A_2B_2. Genotypes are formed by the union of maternal and paternal gametes.

If an individual is homozygous at one or both loci there is no ambiguity about the types of gametes that the individual received. For example, an $A_1A_1B_1B_2$ must have been formed from the union of A_1B_1 and A_1B_2 gametes, although it will not be known which one came from which parent. The gametic origins of the genotype can be emphasized by writing the genotype with a slash separating the gametes: A_1B_1/A_1B_2. All eight genotypes that are homozygous at one or both loci can be represented by one such gametic arrangement.

There is ambiguity with double heterozygotes: $A_1A_2B_1B_2$. These genotypes arise from two different gametic pairings: A_1B_1/A_2B_2 and A_1B_2/A_2B_1, as indicated in the middle element of the array in Table 4.9. It is not possible to distinguish these two only on the basis of the individual's genotype.

Linkage

For the transmission of gametes, there are three cases to consider. Doubly homozygous individuals received two copies of the same gamete and can transmit only that kind: $A_1A_1B_1B_1$ can transmit only A_1B_1 for example. Individuals who are homozygous at one locus and heterozygous at the other have received two different gametes, and can transmit each of these two with equal probability: $A_1A_1B_1B_2$ individuals transmit A_1B_1 and A_1B_2 equally often.

Because of the possibility of *recombination* between loci, however, doubly heterozygous individuals can transmit four kinds of gametes–the two *parental* types that the individual received and the two *recombinant* types that are different from both parental types. We write the probability of recombination as c and note that each recombinant gamete is transmitted with the same probability of $c/2$ and each parental gamete is transmitted with probability $(1 - c)/2$. Loci that are on different chromosomes, or that are far apart on the same chromosome, are said to be *unlinked*. For such loci, double heterozygotes produce all four types of gamete with equal probability. This corresponds to $c = 0.5$, and the mechanism by which recombination takes place usually ensures that $0 \leq c \leq 0.5$ for all pairs of loci. Loci that are (virtually) at the same position on a chromosome are said to be *completely linked* and have zero recombination fraction, $c = 0$. For such loci, double heterozygotes can transmit only the two parental gametes–each with a probability of 0.5. To summarize this discussion, we show gametic probabilities in Table 4.9.

Linkage disequilibrium

Partly because of the linkage phenomenon, the probability with which an individual receives a specific **A** allele may be related to the probability with which it receives a specific allele at locus **B**. This phenomenon is known as *linkage disequilibrium*, even though it can exist between the alleles at unlinked loci. Formally, the coefficient D_{AB} of linkage disequilibrium for alleles A and B at loci **A** and **B** is defined as the difference between the AB gamete proportion and the product of A and B allele proportions p_A, p_B:

$$D_{AB} = P_{AB} - p_A p_B$$

Even though the two alleles may have proportions that do not change over time, the proportion of the pair AB will change. In Box 4.8 we show how linkage

Table 4.10   Proportions of two-locus genotypes.

	Locus B			
Locus A	$B_1 B_1$	$B_1 B_2$	$B_2 B_2$	Total
$A_1 A_1$	$1 : P_{A_1 B_1}^{A_1 B_1}$	$2 : P_{A_1 B_2}^{A_1 B_1}$	$3 : P_{A_1 B_2}^{A_1 B_2}$	$P_{A_1 A_1}$
$A_1 A_2$	$4 : P_{A_2 B_1}^{A_1 B_1}$	$5 : P_{A_2 B_1}^{A_1 B_2} + P_{A_2 B_2}^{A_1 B_1}$	$6 : P_{A_2 B_2}^{A_1 B_2}$	$P_{A_1 A_2}$
$A_2 A_2$	$7 : P_{A_2 B_1}^{A_2 B_1}$	$8 : P_{A_2 B_2}^{A_2 B_1}$	$9 : P_{A_2 B_2}^{A_2 B_2}$	$P_{A_2 A_2}$
Total	$P_{B_1 B_1}$	$P_{B_1 B_2}$	$P_{B_2 B_2}$	1

disequilibrium decays as recombination rearranges pairs of alleles at different loci. The value D_{AB} in one generation changes to D'_{AB} in the next:

$$D'_{AB} = (1 - c) D_{AB}$$

The linkage disequilibrium decays by a maximum amount of one-half each generation for unlinked genes, and this rate is quite fast. The derivation in Box 4.8 makes use of the notation displayed in Table 4.10 for two-locus genotype proportions.

Disequilibrium in Admixed Populations

One way in which linkage disequilibrium can be created is by the amalgamation of two populations, in the same way that the Wahlund effect generates Hardy-Weinberg disequilibrium. Even if two subpopulations are each in linkage equilibrium there may be linkage disequilibrium in the combined population. The amount of disequilibrium is proportional to the differences between subpopulations of allele proportions at the two loci. If the two subpopulations are indexed by α and β, and allele proportions are $p_{A\alpha}, p_{A\beta}$ for allele A and $p_{B\alpha}, p_{B\beta}$ for allele B, then in the whole population the disequilibrium is

$$D_{AB} = m_\alpha m_\beta [p_{A\alpha} - p_{A\beta}][p_{B\alpha} - p_{B\beta}]$$

where m_α and m_β are the proportions of the whole population in the two subpopulations. This result is derived in Box 4.9.

Table 5.1   An AmpliType^R profile.

Locus	Genotype
LDLR	*AB*
GYPA	*BB*
HBGG	*BC*
D7S8	*AB*
Gc	*BC*

proportions of these genotypes will vary considerably, with some being relatively common and many being very rare. The simplest estimate of the proportion of any particular genotype in a population is just its proportion in a sample from that population, and this is the maximum likelihood estimate discussed in Chapter 3. Unless the sample is extremely large, however, most of these genotypes will not be found in the sample even if they are present in the population.

We can explain this last point as follows. Consider any one of the genotypes; as we have seen in Chapter 3, the number of individuals in a sample with that particular genotype has a binomial probability distribution. The two parameters of the distribution are the known sample size n and the unknown population proportion P of people with that type. We first use the binomial distribution to calculate the probability of there being *no* individuals in the sample with the genotype in question:

$$\Pr(\text{Zero copies of genotype in the sample}) = \frac{n!}{0!n!}P^0(1-P)^n$$
$$= (1-P)^n$$

From this we can calculate the probability of there being one or more copies of this genotype in the sample:

$$\Pr(\text{At least one copy of the genotype in the sample}) = 1-(1-P)^n$$

In Table 5.2 we show, for different values of P, the approximate sample size required for this probability to be at least 0.95. Evidently, databases of a few hundred, or even a few thousand, are not going to ensure that rare genotypes are seen.

The solution to the problem of estimating small multilocus genotype proportions rests on the assumption of independence between the constituent parts of

Table 5.2   Approximate sample sizes n needed to have 0.95 probability of detecting a genotype for which the proportion is P.

P	n
1	1
0.1	30
0.01	300
0.001	3,000
0.000,1	30,000
0.000,01	300,000
0.000,001	3,000,000

the profile. If it is reasonable to assume independence between loci, a multilocus genotype proportion can be estimated reliably by multiplying together the constituent single-locus genotype proportions.

To illustrate the process, we use data collected by Cellmark Diagnostics from a sample of 103 people who were typed at the Polymarker loci. Table 5.3 is in five parts, each one corresponding to the locus shown in column 1. The second column shows the designations of the single-locus genotypes, and the numbers of people in the sample who have that genotype are shown in the third column. These counts are divided by the total of 103 to give the proportions shown in the fourth column. We know these sample proportions are estimates of the corresponding population proportions, and we can gain some idea of their precision by estimating their standard deviations. In Chapter 3 we saw that an estimated proportion $\hat{P}$, based on a sample of size n, has a probability distribution with a standard deviation of $\sqrt{P(1-P)/n}$, where P is the population proportion. We cannot use this equation directly because we don't know P, but statistical theory tells us that replacing P by $\hat{P}$ provides a good estimate of the standard deviation. These estimates, $\sqrt{\hat{P}(1-\hat{P})/n}$, are shown in the fifth column of Table 5.3.

The next thing we can do is recover the allele counts at each locus by the simple counting method described in Chapter 4. If these counts are divided by the total number of alleles, 2×103, we obtain sample allele proportions, and these, in turn, are estimates of the population proportions. If we denote estimates of the proportions for alleles A and B in a system such as $LDLR$ by $\hat{p}_A$, and $\hat{p}_B$, and if the three estimated genotype proportions are $\hat{P}_{AA}, \hat{P}_{AB}$, and $\hat{P}_{BB}$, then we have

$$\hat{p}_A \;\; = \;\; \hat{P}_{AA} + \frac{1}{2}\hat{P}_{AB}$$

Table 5.3   Sample genotype proportions $\hat{P}$ (and standard deviations) for Poly-marker loci.

Locus	Genotype	Count	Observed values $\hat{P}$	(Std. dev.)	Product estimates $\hat{P}$	(Std. dev.)
LDLR	AA	17	0.165	(0.037)	0.191	(0.030)
	AB	56	0.544	(0.049)	0.492	(0.009)
	BB	30	0.291	(0.045)	0.317	(0.039)
GYPA	AA	31	0.301	(0.045)	0.290	(0.037)
	AB	49	0.476	(0.049)	0.497	(0.005)
	BB	23	0.223	(0.041)	0.213	(0.032)
HBGG	AA	30	0.291	(0.045)	0.312	(0.039)
	AB	55	0.534	(0.049)	0.488	(0.010)
	AC	0	0.000	(0.000)	0.005	(0.005)
	BB	17	0.165	(0.037)	0.191	(0.030)
	BC	1	0.010	(0.010)	0.004	(0.004)
	CC	0	0.000	(0.000)	0.000	(0.000)
D7S8	AA	31	0.301	(0.045)	0.296	(0.038)
	AB	50	0.485	(0.049)	0.496	(0.006)
	BB	22	0.214	(0.040)	0.208	(0.032)
Gc	AA	4	0.039	(0.019)	0.064	(0.015)
	AB	11	0.107	(0.030)	0.100	(0.016)
	AC	33	0.320	(0.046)	0.277	(0.026)
	BB	8	0.078	(0.026)	0.040	(0.011)
	BC	14	0.136	(0.034)	0.218	(0.026)
	CC	33	0.320	(0.046)	0.301	(0.038)

Source: Cellmark Diagnostics.

$$\hat{p}_B = \hat{P}_{BB} + \frac{1}{2}\hat{P}_{AB}$$

Allele proportion estimates are shown in Table 5.4, and once again the precision of the estimates can be gauged by the estimated standard deviations also displayed in this table. Theoretical expressions for these standard deviations are derived in Box 5.1.

Box 5.1: Standard deviation of estimated allele proportions

Allele proportions are linear combinations of genotype proportions. If the sample proportion of allele A_i is $\hat{p}_i$, and the sample proportion of genotype $A_i A_j$ is $\hat{P}_{ij}$, then the variance of $\hat{p}_i$ is

$$
\begin{aligned}
\mathrm{Var}(\hat{p}_i) &= \mathrm{Var}\!\left(\hat{P}_{ii} + \frac{1}{2}\sum_{j \neq i}\hat{P}_{ij}\right) \\[2mm]
&= \mathrm{Var}(\hat{P}_{ii}) + \sum_{j \neq i}\mathrm{Cov}(\hat{P}_{ii}, \hat{P}_{ij}) + \frac{1}{4}\sum_{j \neq i}\mathrm{Var}(\hat{P}_{ij}) \\[2mm]
&= \frac{1}{2n}(p_i + P_{ii} - 2p_i^2)
\end{aligned}
$$

where Cov denotes covariance, that is, the expected value of the product of two quantities minus the product of their expected values. When the Hardy-Weinberg relationship holds, so that $P_{ii} = p_i^2$, this reduces to

$$
\mathrm{Var}(\hat{p}_i) = \frac{1}{2n}p_i(1 - p_i)
$$

as expected for a sample from a binomial distribution $B(2n, p_i)$.

THE PRODUCT RULE

So far we have made no assumptions about the conditions for independence of alleles within and between loci. If we can assume that the conditions for independence of alleles within loci exist in the sampled population, then we can use the Hardy-Weinberg formula to estimate population genotype proportions from the estimated allele proportions:

$$
\left.
\begin{aligned}
P_{AA} &\;\hat{=}\; \hat{p}_A^2 \\
P_{AB} &\;\hat{=}\; 2\hat{p}_A\hat{p}_B \\
P_{BB} &\;\hat{=}\; \hat{p}_B^2
\end{aligned}
\right\}
\tag{5.1}
$$

Here the symbol $\hat{=}$ means "is estimated by." These product estimates are displayed in column 6 of Table 5.3.

In the last column of Table 5.3 we give estimated standard deviations for the genotype proportions estimated as the products of allele proportions. The standard deviations are calculated by a method described in Box 5.2. Table 5.3

this *LDLR* sample, the test statistic is much less than 3.84, and we do not reject the Hardy-Weinberg hypothesis.

It is important to note that the test must be performed on counts, not on proportions or percentages. Suppose the data were 90 *AA*, 0 *AB*, and 10 *BB*. The absence of *AB* genotypes is a strong indication of departures from Hardy-Weinberg equilibrium, and the test statistic has the value of 100:

$$X^2 = \frac{(90 - 81.0)^2}{81.0} + \frac{(0 - 18.0)^2}{18.0} + \frac{(10 - 1.0)^2}{1.0} = 100$$

If the statistic had been calculated incorrectly with proportions (0.90, 0.00, 0.10 observed and 0.81, 0.18, 0.01 expected) rather than counts, however, it would have the value of 1 and would not lead to rejection of the Hardy-Weinberg hypothesis.

We explained in Chapter 3 that the chi-square goodness-of-fit test becomes unreliable when one or more of the expected counts are small. It is conventional to require that all expected counts are at least five, although this ad-hoc rule can be relaxed. For loci with many alleles, however, it is common to find small expected counts even for large sample sizes, and for this reason we prefer to use exact tests.

Exercise 5.1   Perform goodness-of-fit tests for Hardy-Weinberg for: (a) *GYPA*; (b) *HBGG*; (c) *D7S8*; and (d) *Gc* using the data in Table 5.3.

Exact Tests.   In Chapter 3 we explained that the exact test is based on the idea of using the multinomial distribution to calculate the probability of the observed data given that the null hypothesis is true. The hypothesis is rejected if this probability belongs to the set of smallest (5%) of possible values. This is in contrast to the goodness-of-fit test, which rejects the hypothesis when the test statistic is larger than expected if the hypothesis is true.

For a locus such as *LDLR* with two alleles *A* and *B*, the probability needed is that of the genotype counts in the sample conditional on the allele counts and conditional on the Hardy-Weinberg hypothesis being true. In Box 5.7 we derive this probability:

$$\Pr(n_{AA}, n_{AB}, n_{BB} | n_A, n_B) = \frac{n! n_A! n_B! 2^{n_{AB}}}{(2n)! n_{AA}! n_{AB}! n_{BB}!} \tag{5.4}$$

This equation looks a little forbidding at first sight, and it is usually evaluated by computer, but we can illustrate its use for the *LDLR* data in Table 5.3:

$$\Pr(n_{AA}, n_{AB}, n_{BB} | n_A, n_B) = \frac{103! 90! 116! 2^{56}}{206! 17! 56! 30!} = 0.0958$$

For loci with more than two alleles, there is a natural extension of Equation 5.4. We index the genotypes by g and the alleles by a, and write H for the number of individuals in the sample who are heterozygous at this locus. The equation becomes

$$\Pr(\{n_g\}|\{n_a\}) = \frac{n!2^H \prod_a n_a!}{(2n)! \prod_g n_g!}$$

where $\{n_g\}$ refers to the collection of genotype counts and $\{n_a\}$ refers to the collection of allele counts.

At this point we must be careful to avoid a possible source of confusion. When we performed the goodness-of-fit test we referred to the chi-square distribution to find the probability of the calculated test statistic *or a greater value* if the null hypothesis was true. This tail area probability is also called the significance level. It corresponds to unlikely values if the hypothesis is true, and these unlikely values are the *largest* values. When we perform the exact test the test statistic itself is a probability, but the statistic does not give the tail probability. The tail probability is the sum of the probabilities for all the outcomes that are as probable as or less probable than the observed outcome, i.e., we add together the smallest probabilities.

How do we determine if the exact-test probability indicates that the sample belongs to the set of unusual outcomes if the Hardy-Weinberg hypothesis is true? We need the total probability of all the possible outcomes that are as probable as or are less probable than the observed outcome. This total is the significance level. We can either quote this value for any set of genotypes, or we can say the hypothesis is rejected at the $\alpha\%$ level if the significance level is less than α. Because the observed outcome belongs to the set of outcomes that determine the significance level, its probability is less than (or equal to) the significance level. In our *LDLR* example the probability of the data is already bigger than the conventional 0.05 level for hypothesis rejection, so the significance level must also be bigger than 0.05.

All 46 possible sets of genotypes for $n_A = 90, n_B = 116$ are shown in Table 5.7, along with their probabilities calculated from Equation 5.4, which assumes the Hardy-Weinberg hypothesis is true. Notice only five of the 46 have probabilities greater than the 0.0958 value for our data. The sum of the probabilities for the remaining 41 sets (which includes the data) is 0.3239. So, for these data, the significance level or P-value is 0.3239.

Permutation-Based Significance Levels

For a locus with only two alleles, it is not difficult to examine all possible genotypic arrays for a given allelic array as shown in Table 5.7, particularly when the

Box 5.7: Probabilities needed for the exact test of the Hardy-Weinberg hypothesis

Recall that the multinomial probability for genotype counts n_{AA}, n_{AB}, and n_{BB} for a locus with alleles A and B is

$$\Pr(n_{AA}, n_{AB}, n_{BB}) = \frac{n!}{n_{AA}! n_{AB}! n_{BB}!} (p_{AA})^{n_{AA}} (p_{AB})^{n_{AB}} (p_{BB})^{n_{BB}}$$

Under the Hardy-Weinberg hypothesis, genotype proportions are replaced by products of allele proportions, and the probability can be written as

$$\Pr(n_{AA}, n_{AB}, n_{BB}, n_A, n_B) = \frac{n!}{n_{AA}! n_{AB}! n_{BB}!} 2^{n_{AB}} (p_A)^{n_A} (p_B)^{n_B} \qquad (5.5)$$

where n_A and n_B are the two allele counts. The difficulty with this expression is that we don't know the population allele proportions. One way around this is to work with the probability of the genotype counts conditional on the allele counts. We ask whether the arrangement of n_A alleles of type A and n_B alleles of type B into genotype counts n_{AA}, n_{AB}, and n_{BB} falls among the most unlikely arrangements if the Hardy-Weinberg relation holds. This has been found to lead to satisfactory Hardy-Weinberg tests (Maiste and Weir 1995).
From the third law of probability,

$$\Pr(n_{AA}, n_{AB}, n_{BB} | n_A, n_B) = \frac{\Pr(n_{AA}, n_{AB}, n_{BB}, n_A, n_B)}{\Pr(n_A, n_B)} \qquad (5.6)$$

What is the probability of the allele counts? Under the Hardy-Weinberg hypothesis, alleles are independent, so a sample of n genotypes is equivalent to a sample of $2n$ alleles. The binomial distribution holds for the two alleles:

$$\Pr(n_A, n_B) = \frac{(2n)!}{n_A! n_B!} (p_A)^{n_A} (p_B)^{n_B} \qquad (5.7)$$

Putting Equations 5.5, 5.6, and 5.7 together provides the conditional probability of the genotype counts if Hardy-Weinberg holds:

$$\Pr(n_{AA}, n_{AB}, n_{BB} | n_A, n_B) = \frac{n! n_A! n_B! 2^{n_{AB}}}{(2n)! n_{AA}! n_{AB}! n_{BB}!}$$

When this probability is used for an exact test with permutation-based significance levels, the value obtained for a data set is compared to values obtained for data sets formed by permuting alleles among genotypes. In each of the permuted sets, the samples sizes $(n, 2n)$ remain the same, as do the allele counts (n_a). Comparisons can therefore be restricted to the ratio $2^{n_{AB}}/(n_{AA}! n_{AB}! n_{BB}!)$.

Table 5.7   All possible samples with $n_A = 90, n_B = 116$, together with probabilities calculated assuming the Hardy-Weinberg hypothesis is true.

n_{AA}	n_{AB}	n_{BB}	Prob.	Cum. prob.	n_{AA}	n_{AB}	n_{BB}	Prob.	Cum. prob.
45	0	58	0.0000	0.0000	31	28	44	0.0000	0.0000
44	2	57	0.0000	0.0000	9	72	22	0.0000	0.0000
43	4	56	0.0000	0.0000	30	30	43	0.0000	0.0000
42	6	55	0.0000	0.0000	10	70	23	0.0001	0.0001
41	8	54	0.0000	0.0000	29	32	42	0.0001	0.0003
40	10	53	0.0000	0.0000	11	68	24	0.0004	0.0007
0	90	13	0.0000	0.0000	28	34	41	0.0005	0.0012
39	12	52	0.0000	0.0000	12	66	25	0.0016	0.0028
1	88	14	0.0000	0.0000	27	36	40	0.0019	0.0047
38	14	51	0.0000	0.0000	13	64	26	0.0052	0.0098
2	86	15	0.0000	0.0000	26	38	39	0.0057	0.0155
37	16	50	0.0000	0.0000	14	62	27	0.0138	0.0293
3	84	16	0.0000	0.0000	25	40	38	0.0148	0.0441
36	18	49	0.0000	0.0000	15	60	28	0.0310	0.0751
4	82	17	0.0000	0.0000	24	42	37	0.0327	0.1078
35	20	48	0.0000	0.0000	16	58	29	0.0591	0.1668
5	80	18	0.0000	0.0000	23	44	36	0.0613	0.2282
34	22	47	0.0000	0.0000	17	56	30	0.0958	0.3239
6	78	19	0.0000	0.0000	22	46	35	0.0982	0.4221
33	24	46	0.0000	0.0000	18	54	31	0.1321	0.5542
7	76	20	0.0000	0.0000	21	48	34	0.1340	0.6882
32	26	45	0.0000	0.0000	19	52	32	0.1555	0.8438
8	74	21	0.0000	0.0000	20	50	33	0.1562	1.0000

sample sizes are not too large. For loci with many alleles, however, the number of genotypic arrays is too large to handle even on a computer. In those situations we employ *permutation* procedures. Instead of examining all possible genotypic arrays, we choose a random sample of the arrays by permuting (or shuffling) the alleles.

The process can be visualized as one of constructing a deck of cards, one for each of the n individuals in the sample. One side of each card is labeled with the two alleles observed for that individual, and then the card is torn in half between the two labels. This results in a deck of $2n$ cards, each card now showing just one allele label. Tearing the cards corresponds to breaking whatever

Chapter 6

Parentage Testing

INTRODUCTION

When we considered transfer evidence in Chapter 2, we assumed the framework of a criminal trial in which the two propositions H_p and H_d were to be considered. The subscripts referred to "prosecution" and "defense." In this chapter we will consider parentage disputes, which usually result in civil proceedings. However, we find it convenient to keep the same subscripts. For a civil trial, the plaintiff's proposition H_p will generally be the allegation of a woman that the defendant is the father of her child. Proposition H_d is still that of the defendant, and this may simply be that he is not the father. The notation has the convenient feature that it applies when parentage disputes result in criminal trials, as in cases of rape or incest. Throughout the chapter we use the terms "mother" and "father" to mean biological parents.

For a paternity case we write M for the mother of child C, and AF for the alleged father. Their genotypes will be denoted G_M, G_C, and G_{AF}, respectively. The two propositions are

> H_p: AF is the father of C.
> H_d: Some other man is the father of C.

Later we will consider more complex parentage analyses, often involving members of the same family. These arise in some cases of incest and in the identification of human remains.

EVALUATION OF EVIDENCE

As in Chapter 2 we use E to summarize all the genetic evidence: the genotypes G_M, G_C, and G_{AF} of mother, child and alleged father. We use I for the non-genetic evidence, and this could include statements made by the mother and the alleged father about their relationship. Using Bayes' theorem, our interpretation of the evidence is

$$\frac{\Pr(H_p|E, I)}{\Pr(H_d|E, I)} = \frac{\Pr(E|H_p, I)}{\Pr(E|H_d, I)} \times \frac{\Pr(H_p|I)}{\Pr(H_d|I)} \tag{6.1}$$

or

$$\text{Posterior odds} = \text{Likelihood ratio} \times \text{Prior odds}$$

We direct our attention to evaluation of the likelihood ratio, LR, as we do throughout the book, but we need first to mention three terms that are used in the field of parentage testing (Walker et al. 1983). The first term is the *paternity index* (PI), which is simply another name for the likelihood ratio in Equation 6.1. In simple paternity cases, the terms LR and PI are interchangeable. The second term is *probability of paternity*, meaning the posterior probability of paternity, and the third is the *probability of exclusion* to which we return later in the chapter.

For the probability of paternity we observe that

$$\Pr(H_d|E, I) = 1 - \Pr(H_p|E, I)$$
$$\Pr(H_d|I) = 1 - \Pr(H_p|I)$$

so that Equation 6.1 can be rewritten in terms of posterior and prior probabilities of H_p, and rearranged to give

$$\Pr(H_p|E, I) = \frac{\text{LR} \times \Pr(H_p|I)}{\text{LR} \times \Pr(H_p|I) + [1 - \Pr(H_p|I)]}$$

If the prior odds are one, meaning that the prior probability of paternity is 0.5, the posterior probability of paternity is

$$\Pr(H_p|E, I) = \frac{\text{LR}}{\text{LR} + 1}$$

and this is the quantity that is referred to as the probability of paternity. The derivation of this is due to Essen-Möller (1938). We do not advocate the use of this probability of paternity because of the implicit assumption of a prior probability of 0.5, irrespective of the nongenetic evidence. We will not discuss the issue further, but refer the reader to the excellent discussion by Kaye (1990). The assumption of 50% prior probability is difficult to defend. At the least, any presentation of probabilities of paternity should be accompanied by a table, such as Table 6.1, showing the effects of different prior probabilities.

Table 6.6 LR values for the alternative propositions that the alleged father either is or is not the father when mother, father, and alleged father all belong to the same subpopulation. (Different subscripts indicate different alleles.) The proportions p_i refer to the whole population.

G_M	G_C	A_M	A_P	G_{AF}	PI	PI when $\theta = 0.03, p_i = 0.1$
A_iA_i	A_iA_i	A_i	A_i	A_iA_i	$\dfrac{1+3\theta}{4\theta+(1-\theta)p_i}$	5.0
				A_iA_j	$\dfrac{1+3\theta}{2[3\theta+(1-\theta)p_i]}$	3.0
	A_iA_j	A_i	A_j	A_jA_j	$\dfrac{1+3\theta}{2\theta+(1-\theta)p_j}$	6.6
				A_iA_j	$\dfrac{1+3\theta}{2[\theta+(1-\theta)p_j]}$	4.5
				A_jA_k	$\dfrac{1+3\theta}{2[\theta+(1-\theta)p_j]}$	4.5
A_iA_k	A_iA_i	A_i	A_i	A_iA_i	$\dfrac{1+3\theta}{3\theta+(1-\theta)p_i}$	6.0
				A_iA_k	$\dfrac{1+3\theta}{2[2\theta+(1-\theta)p_i]}$	3.6
	A_iA_j	A_i	A_j	A_jA_j	$\dfrac{1+3\theta}{2\theta+(1-\theta)p_j}$	6.6
				A_iA_j	$\dfrac{1+3\theta}{2[\theta+(1-\theta)p_j]}$	4.5
				A_jA_l	$\dfrac{1+3\theta}{2[\theta+(1-\theta)p_j]}$	4.5

PATERNITY EXCLUSION

In paternity disputes the question is whether or not a particular man is the father of a particular child. Classical considerations of such questions were limited to excluding a man from paternity of a child when the man did not have the child's paternal allele at some locus, or, if the paternal allele cannot be determined, when

Table 6.7   Paternity exclusion configurations at one locus with an arbitrary number of alleles; k is any value different from i and j.

Mother		Child		Excluded man	
Type	Probability	Type	Probability[1]	Genotypes	Probability
A_iA_i	p_i^2	A_iA_i	p_i	$A_wA_x,\ w,x \neq i$	$(1-p_i)^2$
		A_iA_j	p_j	$A_wA_x,\ w,x \neq j$	$(1-p_j)^2$
A_iA_j	$2p_ip_j$	A_iA_i	$p_i/2$	$A_wA_x,\ w,x \neq i$	$(1-p_i)^2$
$j \neq i$		A_jA_j	$p_j/2$	$A_wA_x,\ w,x \neq j$	$(1-p_j)^2$
		A_iA_j	$(p_i+p_j)/2$	$A_wA_x,\ w,x \neq i,j$	$(1-p_i-p_j)^2$
		A_iA_k	$p_k/2$	$A_wA_x,\ w,x \neq k$	$(1-p_k)^2$
		A_jA_k	$p_k/2$	$A_wA_x,\ w,x \neq k$	$(1-p_k)^2$

[1]Probability of genotype of child given genotype of mother.

the man had neither of the child's alleles. The increasing availability of diagnostic loci has given rise to calculations based on the probabilities of the child's genotype under alternative propositions as we have shown. For completeness, however, we now review some results pertaining to exclusion.

In Table 6.7 we show all the possible combinations of genotypes of a mother and her child for a locus with alleles A_i. The last column of the table shows the probability that an unknown man will be excluded as being the father of the child. This calculation has nothing to do with any specific alleged father, and it is seen to depend on the population proportion of the allele inferred to be the child's paternal allele. Except for the case where both mother and child have the same heterozygous genotype, the paternal allele is easily identified as the allele the child did not get from the mother. For the double heterozygote case, either allele could have been the paternal allele, so the exclusion probability uses the sum of the proportions of both alleles. If there are only two alleles, no man could be excluded from paternity by this locus.

A genetic marker can be characterized by its ability to exclude an unrelated man from paternity in any situation. This exclusion probability is given by summing the joint probabilities of all the mother-child-excluded man combinations shown in the table. The probability of an A_iA_i mother with an A_iA_i child is $p_i^2 \times p_i$, and this combination excludes all men that do not have an A_i allele. Such men occur in proportion $(1 - p_i)^2$, so that the trio in the first line of Taso that

$$\text{LR} = \frac{1}{4 p_1 p_3}$$

Exercise 6.4   Find the likelihood ratio for the situation when profiles are available from the mother, four siblings, the spouse, and a child of a missing person, as well as from a sample that may be from the missing person. Specifically,

$$
\begin{array}{ll}
\text{Mother}: & G_P = A_3 A_4 \\
\text{Sibs}: & \{G_S\} = \{A_2 A_4, A_2 A_4, A_2 A_4, A_3 A_4\} \\
\text{Spouse}: & G_M = A_5 A_6 \\
\text{Child}: & G_C = A_3 A_5 \\
\text{Sample}: & G_X = A_3 A_3
\end{array}
$$

Use the fact that the genotypes of the four siblings and the mother imply that the father of the missing person must have had genotype $A_2 A_3$ or $A_2 A_4$.

Deceased Alleged Father

A similar scenario is when the alleged father in a paternity dispute cannot be typed, but typing is available from his relative(s). Suppose profiles are available from the mother and her child, allowing the paternal allele $A_P = A_i$ to be determined. The alleged father AF is deceased but his relative Z has been typed.

We need the probability that AF would transmit allele A_i when Z has a specified genotype. This is another instance where three-allele probabilities are needed. From the previous results for non-inbred individuals:

$$
\begin{array}{ll}
\Pr(A_P = A_i | G_z = A_i A_i) &= 2\theta_{AF,Z} + (1 - 2\theta_{AF,Z}) p_i \\
\Pr(A_P = A_i | G_z = A_i A_j) &= \theta_{AF,Z} + (1 - 2\theta_{AF,Z}) p_i, \ \ j \neq i \\
\Pr(A_P = A_i | G_z = A_j A_k) &= (1 - 2\theta_{AF,Z}) p_i, \ \ j, k \neq i
\end{array}
$$

so the likelihood ratio is

$$
\begin{aligned}
\text{LR} &= \frac{\Pr(A_M | G_M, H_p) \Pr(A_P | G_Z, H_p) \Pr(G_Z | H_p)}{\Pr(A_M | G_M, H_d) \Pr(A_P | G_Z, H_d) \Pr(G_Z | H_d)} \\[2mm]
&= \frac{\Pr(A_P | G_Z, H_p)}{\Pr(A_P | H_d)} \\[2mm]
&= \begin{cases}
(1 - 2\theta_{AF,Z}) + \dfrac{2\theta_{AF,Z}}{p_i} & \text{if } G_Z = A_i A_i \\[2mm]
(1 - 2\theta_{AF,Z}) + \dfrac{\theta_{AF,Z}}{p_i} & \text{if } G_Z = A_i A_j, \ j \neq i \\[2mm]
(1 - 2\theta_{AF,Z}) & \text{if } G_Z = A_j A_k, \ j, k \neq i
\end{cases}
\end{aligned}
$$

Table 6.9   LR values for an inheritance dispute.

G_X, G_Y	$\Pr(G_X, G_Y \| H_p)$	$\Pr(G_X \| H_d) \Pr(G_Y \| H_d)$	LR
$A_i A_i, A_i A_i$	$\frac{1}{2} p_i^3 (1 + p_i)$	p_i^4	$\frac{1}{2} + \frac{1}{2p_i}$
$A_i A_i, A_i A_j$	$\frac{1}{2} p_i^2 p_j (1 + 2p_i)$	$2p_i^3 p_j$	$\frac{1}{2} + \frac{1}{4p_i}$
$A_i A_i, A_j A_j$	$\frac{1}{2} p_i^2 p_j^2$	$p_i^2 p_j^2$	$\frac{1}{2}$
$A_i A_i, A_j A_k$	$p_i^2 p_j p_k$	$2p_i^2 p_j p_k$	$\frac{1}{2}$
$A_i A_j, A_i A_j$	$\frac{1}{2} p_i p_j (4p_i p_j + p_i + p_j)$	$4p_i^2 p_j^2$	$\frac{1}{2} + \frac{1}{8p_i} + \frac{1}{8p_j}$
$A_i A_j, A_i A_k$	$\frac{1}{2} p_i p_j p_k (4p_i + 1)$	$4p_i^2 p_j p_k$	$\frac{1}{2} + \frac{1}{8p_i}$
$A_i A_j, A_k A_l$	$2p_i p_j p_k p_l$	$4p_i p_j p_k p_l$	$\frac{1}{2}$

If AF and Z are not related, $\theta_{AF,Z} = 0$, there is therefore no information in the genotype of Z concerning the paternity of AF, and the LR is 1 (Brenner 1997).

Inheritance Dispute

Brenner (1997) discusses the following inheritance dispute: People X and Y have different mothers. The father Z of X has died, and Y claims also to be a child of Z. The two propositions for the genetic evidence of the genotypes of X and Y are

H_p: X and Y are half sibs.
H_d: X and Y are unrelated.

The likelihood ratio is

$$\text{LR} = \frac{\Pr(G_X, G_Y | H_p)}{\Pr(G_X, G_Y | H_d)}$$

$$= \frac{\Pr(G_X, G_Y | H_p)}{\Pr(G_X | H_d) \Pr(G_Y | H_d)}$$

because the genotypes G_X and G_Y are independent when X and Y are unrelated. We now need to go back to Chapter 4 for the joint probabilities of genotypes of half sibs and list them in column 2 of Table 6.9. The joint probabilities in column 3 of Table 6.9 are just the products of the two separate probabilities. We have assumed no inbreeding and no population structure.

The likelihood ratios in Table 6.9 are less than one and against X and Y being half sibs unless they share rare alleles; for example for genotypes $A_i A_j$, $A_i A_k$ we must have $p_i < 1/8$ for LR> 1, and even then LR can decrease with more loci.

SUMMARY

Parentage testing, identification of remains, and inheritance disputes all exploit the genetic laws of transmission of alleles from parent to child. As with forensic applications, the DNA evidence is interpreted with likelihood ratios that compare the probabilities of the evidence under alternative propositions. Evaluation of the probabilities depends on using the laws of probability to make the probability of a genotype conditional on the parental genotype(s).

Chapter 7

Mixtures

INTRODUCTION

In Chapter 2 we considered the case in which the evidence at the crime scene consisted of two blood stains, and there was a single suspect whose genotype was the same as that of one of the stains. In this chapter we are going to extend that discussion considerably by talking about DNA profiles of samples that contain material from more than one contributor. The sensitivity of modern techniques is such that the incidence, complexity, and importance of such cases are increasing. It is not possible to present an exhaustive treatment of every eventuality, but by considering a range of different kinds of cases we hope to assist the reader in gaining a sufficient depth of understanding to tackle other situations as they arise. Rather than providing a recipe book, we adhere to the three principles for interpreting evidence listed in Chapter 2.

We will begin by considering the case in which independence of alleles within and between loci can reasonably be assumed, there are no population substructuring effects of practical magnitude, and all contributors to the mixed profile are from the same population (Evett et al. 1991). Later in the chapter we will relax these assumptions, but we will always assume that all contributors to the mixed profile are unrelated to each other, and that allelic dropout has no practical impact. This last assumption means that we will not be using the "2p" method that is widely used for the estimation of the frequency of single-banded RFLP-based VNTR systems (Weir et al. 1997). Moreover, we will carry out the analysis ignoring the intensities of electrophoretic bands or typing strip dots, or peak heights in histograms generated by automatic sequencers. We can refer to recent publications that do take into account peak heights (Evett et al. 1998). We adopt a simple illustrative method for summarizing profiles: each allele will be represented by an open rectangle.

In our treatments of transfer evidence so far, we have always been able to

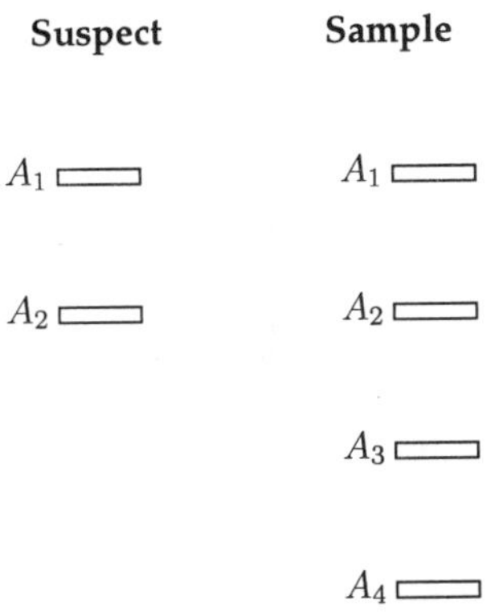

Suspect	Sample
A_1 ▭	A_1 ▭
A_2 ▭	A_2 ▭
	A_3 ▭
	A_4 ▭

Figure 7.4   Four-allele mixture.

SUSPECT AND UNKNOWN PERSON

Some crime samples will contain DNA from more than one person, but only one known person is suspected of being a contributor. As before, we separate the cases of the sample showing three or four alleles.

Four-Allele Mixture

As in Chapter 2, when we considered the case of two stains at the crime scene, the hypotheses for a four-allele mixture profile that includes the genotype of a single suspect (Figure 7.4) are

H_p: The crime sample contains DNA from the suspect and an unknown person.
H_d: The crime sample contains DNA from two unknown people.

Following a similar line of reasoning as in the case where the victim was a contributor, we can show that

$$LR = \frac{\Pr(E_C | G_S, H_p)}{\Pr(E_C | G_S, H_d)}$$

The genotypes G_i that might be components of the crime sample profile are

i	G_i
1	$A_1 A_2$
2	$A_1 A_3$
3	$A_1 A_4$
4	$A_2 A_3$
5	$A_2 A_4$
6	$A_3 A_4$

and we note that $G_1 = G_S$. The law of total probability gives

$$\Pr(E_C|G_S, H_p) = \sum_i \Pr(E_C|G_S, G_i, H_p) \Pr(G_i|G_S, H_p)$$

where G_i denotes the genotype of the unknown contributor to the mixture in addition to the suspect. This simplifies slightly to

$$\Pr(E_C|G_S, H_p) = \sum_i \Pr(E_C|G_1, G_i, H_p) \Pr(G_i|H_p)$$

where we have written G_1 for G_S. However, when we look at the possible values for G_i, we see that the evidence is not possible unless $i = 6$. So $\Pr(E_C|G_1, G_i, H_p) = 0$ when $i \neq 6$. The numerator for the likelihood ratio becomes

$$\begin{aligned}\Pr(E_C|G_S, H_p) &= \Pr(E_C|G_1, G_6, H_p) \Pr(G_6|H_p) \\ &= 1 \times 2p_3 p_4\end{aligned}$$

For the denominator of the likelihood ratio we first invoke our assumption that knowledge of G_S provides no information about the genotypes of possible contributors to the crime sample when H_d is true. Therefore

$$\Pr(E_C|G_S, H_d) = \Pr(E_C|H_d)$$

Because there are two unknown contributors under H_d, with genotypes G_i and G_j, the law of total probability gives

$$\Pr(E_C|H_d) = \sum_i \sum_j \Pr(E_C|G_i, G_j, H_d) \Pr(G_i, G_j, H_d)$$

The double summation indicates that we must consider every possible pair of genotypes from the list of permitted genotypes. Looking at that list, we see that there are only six combinations of i and j for which $\Pr(E_C|G_i, G_j, H_d)$ is not zero. They are all the possible pairs of heterozygotes that contain all four alleles $A_1 A_2 A_3 A_4$ between them:

i	j	G_i	G_j	$\Pr(G_i, G_j \mid H_d)$
1	6	$A_1 A_2$	$A_3 A_4$	$2p_1 p_2 \times 2p_3 p_4$
2	5	$A_1 A_3$	$A_2 A_4$	$2p_1 p_3 \times 2p_2 p_4$
3	4	$A_1 A_4$	$A_2 A_3$	$2p_1 p_4 \times 2p_2 p_3$
4	3	$A_2 A_3$	$A_1 A_4$	$2p_2 p_3 \times 2p_1 p_4$
5	2	$A_2 A_4$	$A_1 A_3$	$2p_2 p_4 \times 2p_1 p_3$
6	1	$A_3 A_4$	$A_1 A_2$	$2p_3 p_4 \times 2p_1 p_2$

We are still assuming that contributors i and j are unrelated, so that G_i and G_j are independent, and that any intensity differences are ignored. For each of the six combinations $\Pr(E_C \mid G_i, G_j, H_d) = 1$, and the likelihood ratio is

$$\begin{aligned} \text{LR} &= \frac{2p_3 p_4}{24 p_1 p_2 p_3 p_4} \\ &= \frac{1}{12 p_1 p_2} \end{aligned}$$

The likelihood ratio is reduced by a factor of six over what it would be if a suspect of type $A_1 A_2$ were included in a single stain of type $A_1 A_2$. If alleles A_1, A_2 were common in the population (e.g. $p_1 = p_2 = 0.3$ giving $12 p_1 p_2 > 1$) the likelihood ratio is actually less than one, meaning that the sample profile is more likely to be of type $A_1 A_2 A_3 A_4$ if it came from two unknown people than if it came from the suspect and one unknown person. The evidence favors the defense, and this is one reason it is important to use the principles of evidence interpretation instead of simplistic rules of the "random man not excluded" type.

Three-Allele Mixture

Suspect heterozygous. When the suspect is heterozygous and the crime sample has those two alleles plus one other, as in Figure 7.5, the likelihood ratio is

$$\text{LR} = \frac{\Pr(E_C \mid G_S, H_p)}{\Pr(E_C \mid G_S, H_d)}$$

and the genotypes that might be components of the crime sample profile are

i	G_i	$\Pr(G_i \mid H_p)$
1	$A_1 A_2$	$2p_1 p_2$
2	$A_1 A_3$	$2p_1 p_3$
3	$A_2 A_3$	$2p_2 p_3$
4	$A_1 A_1$	p_1^2
5	$A_2 A_2$	p_2^2
6	$A_3 A_3$	p_3^2

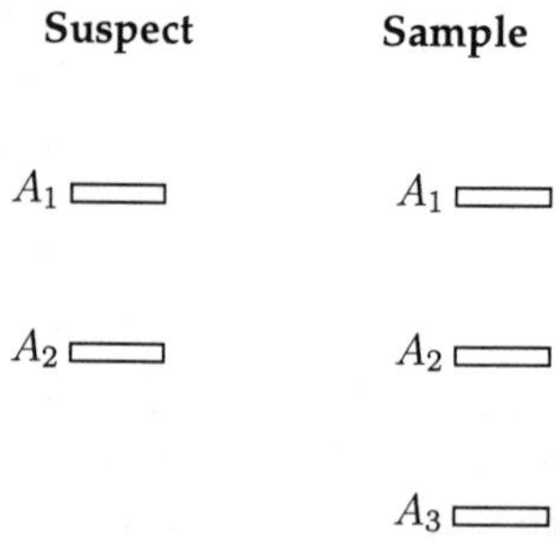

Suspect   **Sample**

$A_1 \rule[0.3ex]{1em}{0.4pt}\!\!\sqsupset$ $A_1 \rule[0.3ex]{1em}{0.4pt}\!\!\sqsupset$

$A_2 \rule[0.3ex]{1em}{0.4pt}\!\!\sqsupset$ $A_2 \rule[0.3ex]{1em}{0.4pt}\!\!\sqsupset$

$A_3 \rule[0.3ex]{1em}{0.4pt}\!\!\sqsupset$

Figure 7.5   Three-allele mixture, suspect heterozygous.

Note that $G_1 = G_S$. The numerator of the likelihood ratio is

$$\Pr(E_C|G_S, H_p) = \sum_i \Pr(E_C|G_S, G_i, H_p)\Pr(G_i|G_S, H_p)$$

$$= \sum_i \Pr(E_C|G_1, G_i, H_p)\Pr(G_i|H_p)$$

Ignoring any intensity differences, only $i = 2, 3, 6$ allow the crime sample to have profile $A_1 A_2 A_3$, and these all give $\Pr(E_C|G_1, G_i, H_p) = 1$. Adding terms

$$\Pr(E_C|G_S, H_p) = 2p_1p_3 + 2p_2p_3 + p_3^2$$

For the denominator, there are 12 combinations of two genotypes that between them have the same profile as the crime sample. These are shown in Table 7.1. The denominator of the likelihood ratio simplifies to

$$\Pr(E_C|G_S, H_d) = 12p_1p_2p_3(p_1 + p_2 + p_3)$$

and the ratio is

$$\text{LR} = \frac{2p_1 + 2p_2 + p_3}{12p_1p_2(p_1 + p_2 + p_3)}$$

Exercise 7.2   For a crime sample of type A_1, A_2, A_3, known to contain DNA from two contributors, evaluate the likelihood ratio in the case where a suspect is of type $A_2 A_2$.

Table 7.2   Notation for mixture calculations.

Alleles in the profile of the evidence sample.

$\mathcal{C}$	The set of alleles in the evidence profile.
$\mathcal{C}_g$	The set of distinct alleles in the evidence profile.
n_C	The known number of contributors to $\mathcal{C}$.
h_C	The unknown number of heterozygous contributors.
c	The known number of distinct alleles in $\mathcal{C}_g$.
c_i	The unknown number of copies of allele A_i in $\mathcal{C}$.
	$1 \leq c_i \leq 2n_C$, $\sum_{i=1}^c c_i = 2n_C$

Alleles from typed people that H declares to be contributors.

$\mathcal{T}$	The set of alleles carried by the declared contributors to $\mathcal{C}$.
$\mathcal{T}_g$	The set of distinct alleles carried by the declared contributors.
n_T	The known number of declared contributors to $\mathcal{C}$.
h_T	The known number of heterozygous declared contributors.
t	The known number of distinct alleles carried by n_T declared contributors.
t_i	The known number of copies of allele A_i in $\mathcal{T}$.
	$0 \leq t_i \leq 2T$, $\sum_{i=1}^s t_i = 2T$.

Alleles from unknown people that H declares to be contributors.

$\mathcal{U}$	The set of alleles carried by the unknown contributors to $\mathcal{C}$.
x	The specified number of unknown contributors to $\mathcal{C}$: $n_C = n_T + x$.
$c - t$	The known number of alleles that are required to be in $\mathcal{U}$.
r	The known number of alleles in $\mathcal{U}$ that can be any allele in $\mathcal{C}_g$, $r = 2x - (c - t)$.
r_i	The unknown number of copies of A_i among the r unconstrained alleles in $\mathcal{U}$.
	$0 \leq r_i \leq r$, $\sum_{i=1}^c r_i = r$.
u_i	The unknown number of copies of A_i in $\mathcal{U}$: $c_i = t_i + u_i$, $\sum_{i=1}^c u_i = 2x$.
	If A_i is in $\mathcal{C}$ but not in $\mathcal{T}$, $u_i = r_i + 1$. If A_i is in $\mathcal{C}$ and also in $\mathcal{T}$, $u_i = r_i$.

Alleles from typed people that H declares to be noncontributors.

$\mathcal{V}$	The set of alleles carried by typed people declared not to be contributors to $\mathcal{C}$.
n_V	The known number of people declared not to be contributors to $\mathcal{C}$.
h_V	The known number of heterozygous declared noncontributors.
v_i	The known number of copies of A_i in $\mathcal{V}$. $\sum_i v_i = 2n_V$.

---

the number of distinct alleles as c. The whole set $\mathcal{C}$ contains c_i copies of allele A_i, $\sum_i c_i = 2n_C$. At present we are assuming allelic independence, so the probability of this set of alleles involves the product of probabilities for each allele in the set: $\prod_i p_i^{c_i}$. To complete the probability calculations we need to know the genotypic composition of the contributors to the sample. For a proposition H in which every contributor is specified, the genotypes are known and we need a factor of 2 for each of the h_C heterozygotes. The probability of the evidence is

$$\Pr(\mathcal{C}|H) \;\; = \;\; 2^{h_C}\prod_i p_i^{c_i}$$

However, at least one of the alternative propositions will involve unknown people, and then we may not know the genotypes of all contributors to the crime sample: we may not know the c_i's or h_C. Instead we have a set of n_T declared contributors, with allele counts t_i and h_T heterozygotes among them. There is also a set of n_V people declared not to be contributors, and these people have allele counts v_i and h_V heterozygotes among them. The x unknown people have u_i copies of allele A_i among them, $\sum_i u_i = 2x$ and $c_i = t_i + u_i$, but we may not know the genotypic composition of the x people. We can say that there are $(2x)!/\prod_i u_i!$ ways of arranging the alleles into genotypes, and this takes care of the factors of 2 for heterozygotes. For the counts u_i, the probability of the evidence is now

$$\Pr(\mathcal{C},\mathcal{V}|H) \;\; = \;\; \frac{2^{h_T+h_V}(2x)!}{\prod_i u_i!}\prod_i p_i^{t_i+u_i}$$

It remains to assign values to the unknown counts u_i. The unknown people are constrained to carry at least one copy of the c distinct alleles in the crime sample that are not among the t distinct alleles carried by the declared contributors to the sample. This accounts for $c-t$ of the $2x$ alleles among the unknowns. Otherwise, they may carry any of the alleles in $\mathcal{C}_g$ but may not carry any allele not in $\mathcal{C}_g$. There are $r = 2x - (c-t)$ alleles in this unconstrained set, for which we write r_i for the number of A_i alleles. Note that r is known but the r_i are not known. For an allele in $\mathcal{C}_g$ but not in $\mathcal{T}_g$ we have that $u_i = r_i + 1$, and for the alleles in both $\mathcal{C}_g$ and $\mathcal{T}_g$ we have $u_i = r_i$. For any other allele, $u_i = 0$. It is a straightforward procedure to have a computer assign values to the r_is:

- Let r_1 take each value in the range $0, 1, \ldots, r$.

- Let r_2 take each value in the range $0, 1, \ldots, r - r_1$.

- $\ldots$

- Let r_{c-1} take each value in the range $0, 1, \ldots, r - r_1 - \ldots - r_{c-2}$.

- Then $r_c = r - r_1 - r_2 - \ldots - r_{c-1}$.

We can write the probability of the distinct alleles $C_g = (A_1, A_2, \ldots, A_c)$ in the crime sample, for a proposition that has a set of alleles $\mathcal{T}$ carried by n_T declared contributors, a set $\mathcal{U}$ carried by x unknown contributors, and a set $\mathcal{V}$ carried by n_V known noncontributors, as

$$P_x(\mathcal{T}, \mathcal{U}, \mathcal{V}|C_g) = \sum_{r_1=0}^{r} \sum_{r_2=0}^{r-r_1} \cdots \sum_{r_{c-1}=0}^{r-r_1-\ldots-r_{c-2}} \frac{2^{h_T+h_V}(2x)!}{\prod_i u_i!} \prod_{i=1}^{c} p_i^{t_i+u_i+v_i}$$

As an example, suppose the crime sample has alleles A_1, A_2, and A_3. Three people have been typed, and found to have the genotypes $A_1 A_1$, $A_2 A_3$, and $A_3 A_3$. Such an example, for locus D1S80, was discussed by Weir et al. (1997). Two alternative propositions are

H_p: The crime sample is from the three typed people.
H_d: The crime sample is from three unknown people.

Under H_p, there are no unknown contributors and no noncontributors so, writing the empty set of alleles from unknown contributors or noncontributors as ϕ,

$$P_0(\mathcal{T} = A_1 A_1 A_2 A_3 A_3 A_3, \mathcal{U} = \phi, \mathcal{V} = \phi | A_1 A_2 A_3) = 12 p_1^2 p_2 p_3^3$$

and this is $\Pr(E|H_p)$, the probability of the evidence under H_p.

Under H_d, there are no declared contributors and there are three unknown contributors. The unknown contributors must carry alleles $A_1 A_2 A_3$, leaving $r = 3$ unconstrained alleles, and $u_i = r_i + 1, i = 1, 2, 3$. There are also three declared noncontributors, who have alleles $A_1 A_2 A_3$ with counts $2, 1, 3$ and one heterozygote. The probability, for $\mathcal{T} = \phi, \mathcal{U} = A_1 A_2 A_3, \mathcal{V} = A_1 A_2 A_3$, is

$$P_3(\phi, \mathcal{U}, \mathcal{V} | A_1 A_2 A_3) = \sum_{r_1=0}^{3} \sum_{r_2=0}^{3-r_1} \frac{2^1 p_1^{r_1+1} p_2^{r_2+1} p_3^{4-r_1-r_2}}{(r_1+1)!(r_2+1)!(4-r_1-r_2)!}$$

There are 10 terms in the summation, corresponding to r_1, r_2, and r_3 values of $(0,0,3)$, $(0,1,2)$, $(0,2,1)$, $(0,3,0)$, $(1,0,2)$, $(1,1,1)$, $(1,2,0)$, $(2,0,1)$, $(2,1,0)$, and $(3,0,0)$. The 10 add to the probability of the evidence under H_d:

$$\begin{aligned}
\Pr(E|H_d) = {} & 360 p_1^3 p_2^2 p_3^4 [p_3^3 + 2 p_2 p_3^2 + 2 p_2^2 p_3 + p_2^3 + 2 p_1 p_3^2 \\
& + 3 p_1 p_2 p_3 + 2 p_1 p_2^2 + 2 p_1^2 p_3 + 2 p_1^2 p_2 + p_1^3]
\end{aligned}$$

and the likelihood ratio is the ratio $\Pr(E|H_p)/\Pr(H_d)$.

Effects of Population Structure

The general approach in the previous section assumed that all the alleles were independent. That can be modified very simply (Curran et al. 1999) to allow for the kinds of dependence imposed by joint membership in the same subpopulation. The Dirichlet theory described in Chapter 4 is appropriate, and Equation 4.23 is needed. Recall that this theory does assume independence of alleles within the subpopulation, but dependence in the whole population. The theory supposes that allele proportions are available only for the whole population. The quantity θ serves to quantify the variation of allele proportions among the subpopulations, and allows the population-wide proportions to be used for a subpopulation. The term $\prod_i p_i^{t_i + u_i + v_i}$ in the general expression of the last section is replaced by

$$\frac{\Gamma(\gamma.)}{\Gamma(\gamma. + 2n_T + 2x + 2n_V)} \prod_{i=1}^{c} \frac{\Gamma(\gamma_i + t_i + u_i + v_i)}{\Gamma(\gamma_i)}$$

with $\gamma_i = (1 - \theta)p_i/\theta$ and $\gamma. = \sum_{i=1}^{c} \gamma_i$ as before.

It is possible to refine this to allow some of the people to be in one subpopulation, and some in another (Curran et al. 1999). This would allow a victim and her attacker to be in different subpopulations, or even different racial groups, but the suspect and attacker to be in the same subpopulation. It is necessary to apply separate Dirichlet moment formulations to each different subpopulation and also to account for the orderings of alleles among individuals separately for each subpopulation. This refinement may not be necessary because assigning all alleles to the same subpopulation maximizes the effects of population substructure, and separate analyses can be made with allele proportions from different racial groups.

NUMBER OF CONTRIBUTORS

The analyses so far have all assumed that the number of contributors to a mixed sample is known. In some cases this will be a reasonable assumption, but in other cases there may be little information about the number of unknown contributors. A complete analysis would allow for different numbers of unknown contributors, each number with its own prior probability. However, these priors are likely to be outside the province of the forensic scientist. An alternative is to provide separate analyses for each of a range of numbers of unknown contributors. By and large, for a locus that has all possible alleles present in the mixture, the probability of the mixture profile increases with the number of contributors. It becomes more probable that a large number of contributors will have all the alleles at a locus between them. The opposite is true for a locus where only some of the possible

were not detected by our test. This may be a consequence of the effects indeed being very small, or it may be a consequence of lack of data, or of the design of the test statistic.

"Proving" dependence.   So, if we have a small P-value we reject the null hypothesis, which seems to suggest that the null hypothesis is false. We appear to have proved that there are dependence effects. The danger here is there has been a school of thought that holds the view that this means the product rule should not be used. But this is not necessarily the case: we must recognize that there is a clear difference between *statistical* significance and *practical* significance. It might be that the quantity of data and the design of the test statistic are such that effects are manifested that, although real, are still far too small to have any noticeable impact on the figures put to a court for interpreting a DNA match. The important point here is that the P-value is not a measure of how well the estimation procedure will work in casework.

Multiple testing.   As we employ more and more loci to aid our discrimination in DNA cases, so the potential numbers of independence tests grow. For example, with a six-locus STR multiplex there are six within-locus tests for each database , and the potential for false rejections will increase above the nominal value of 5% that we believe is the case, for example, when we reject on the basis of a chi-square test statistic exceeding 3.84. The Bonferroni correction (Weir 1996) for multiple tests makes a distinction between "comparison-wise" and "experiment-wise" significance levels. Our discussion on P-values so far has been comparison-wise. Each test, considered singly, has probability P of leading to false rejection. If, for example, we were carrying out six independent tests then the experiment-wise significance level is the proportion of times that one or more of all six tests would lead to a false rejection. If it is desired to keep the experiment-wise rate at level P, then the rate for each individual test needs to be decreased to (approximately) P/N when N tests are performed. In the present example, each of the six tests would be conducted at the $P = 0.008$ level, to give an experiment-wise error rate of 5%. A chi-square statistic would need to be larger than 6.96 instead of 3.84 to indicate rejection. This approach is not really satisfactory, as the six tests are not replicates of the same test. We are really interested in departures from Hardy-Weinberg at each locus individually–it is not that all six single-locus tests are addressing exactly the same issue. The number of tests increases if we also examine sets of three, four, five, and six loci and the potential for false rejections would increase even if perfect independence existed–which it doesn't, as we have seen. Knowing that, what should we do when we get a significant rejection for a particular combination?

Prior knowledge.   A serious weakness of the hypothesis-testing approach is that it does not enable account to be taken of prior knowledge. There may be very good prior reasons that dependence effects in a given population are minor. These could include previous studies at other loci; knowledge that the new loci of interest are evolutionarily neutral; demographic information showing that sub-populations have experienced generations of intermixing; or sociological studies showing that inbreeding levels are small.

Post-hoc rationalization.   A natural consequence of the last two points is the practice, of which there are many examples to be found in the literature (e.g., Evett et al. 1996a) where one or more "significant" results in a study are discounted by the authors as being of no importance. The literature has demonstrated various approaches to doing this: carrying out additional tests based on different test statistics; localizing the genotypic combinations that contribute most to the test failure and demonstrating that the departure has small practical effect; citing other studies in which the effect was not observed; and invoking the fact that multiple testing has been carried out, implementing the Bonferroni inequality to weaken the power of all of them.

The issue of testing for independence across loci is even more complicated, and in Box 5.8 we referred to the great difficulty in performing meaningful tests for allelic independence at several loci.

Why should we employ hypothesis testing?   A good proportion of this book has been devoted to hypothesis testing, and the reader can be forgiven for being puzzled by an apparent *volte-face* by the authors. There are indeed some good reasons for using a significance test, and we point out that tests have been used in many of the scientific advances of the 20th century. For example, new drugs are adopted after rejection of null hypotheses that they have no beneficial effect or that they are no better than existing medications.

With much of existing practice steeped in this approach to scientific inference, it may be difficult to introduce a new system of genetic markers into casework unless it can be shown that a minimum standard battery of tests has been applied. Next, some of the tests are very simple to carry out, can be very useful for a first pass over the data, and can provide an early indication that some hitherto unexpected effect is operating–such as, for example, a technical problem leading to allelic drop-out and consequent excess homozygosity. Finally, we must admit that alternative methods can be mathematically far more complex than most hypothesis tests.

MATCH PROBABILITIES

The need for assuming independence is removed when we proceed as though there is always some degree of association between the genotypes of the suspect and the offender, and calculate the match probability $\Pr(G_C|G_S, H_d, I)$ and thus the LR at Equation 2.3. Our expressions for the match probability in Equations 4.20 allow for dependences due to population structure.

In Chapter 4 we developed a theory for these probabilities when, under H_d, the offender and suspect are related by virtue of being in the same family or by virtue of shared evolutionary history. For our Gotham City example, we can ignore the possibility of family relationships. We do want to consider evolutionary dependence, however, and the most conservative statement we can make about the unknown offender is that he belongs to the same subpopulation as the suspect. Because the convenience sample is from the whole population, our treatment of match probabilities took explicit account of population structure by means of the parameter θ (Equations 4.20):

$$\Pr(G_S = A_i A_i | G_C = A_i A_i) = \frac{[2\theta + (1-\theta)p_i][3\theta + (1-\theta)p_i]}{(1+\theta)(1+2\theta)}$$

$$(8.1)$$

$$\Pr(G_S = A_i A_j | G_C = A_i A_j) = \frac{2[\theta + (1-\theta)p_i][\theta + (1-\theta)p_j]}{(1+\theta)(1+2\theta)}$$

Implicit in the development of these equations is the recognition that population structure produces allelic dependence in the whole population. In other words, use of these equations avoids the need to assume allelic independence in the whole population–indeed dependence is assumed. Use of these equations also avoids the need to specify the subpopulation since the result is expected to hold for any subpopulation. The next issue is that of choosing an appropriate values for θ.

Choosing θ

In our Gotham City example we have just the one convenience sample from the whole population, so how can we decide on a value of θ to use? It is a mistake to be prescriptive, and we would urge scientists to consider each case on its own merits. Guidance can be sought from the literature; demographic studies are useful as well as analyses carried out on genotyping systems other than those in forensic use. We also bear in mind that the forensic scientist will, in general, desire to err on the conservative side.

The parameter θ refers to the relationship of pairs of alleles within a subpopulation relative to that between alleles in different subpopulations. It also serves as a measure of differences among subpopulations. The variance of allele proportions among subpopulations is proportional to θ. The ideal situation would be to have data from different subpopulations in order to estimate the θ values appropriate for each one, as described in Chapter 5. This is not practical, not least because of the difficulty in allocating people to subpopulations. There are two possible solutions. One is to refer to previous studies of human population structure, such as the monumental compilation of Cavalli-Sforza et al. (1994).

Although these authors used different loci, they did study very many populations and we consider their results to be relevant. They reported θ estimates that were generally less than 0.05, which is in agreement with our understanding of human evolution and the graph we showed in Figure 4.7 for $N = 100,000$. The other solution is to adopt an arbitrary value of θ that could be considered conservative, as did the 1996 NRC report (National Research Council 1996). The report contains a useful discussion of this issue, and for STR systems, suggests values in the range 0.01 to 0.03 until practitioners acquire the appropriate data to carry out studies of structuring within their own environment. We support this recommendation.

It is worth stressing that we do not attempt to define a point at which the relationship among people gives a θ of zero. Our understanding of human evolution (e.g., Cavalli-Sforza 1998) is that *all humans are related*. This is a simple consequence of the fact that no two people currently alive can have had distinct sets of ancestors for all of the past 200,000 years. On the other hand, of course, it is for small subpopulations that people have had the longest shared history and so have the largest θ values. It seems appropriate to adopt a conservatively high value of θ with the understanding that it would accommodate people of the same race in the same subpopulation as well as people in quite distinct races.

We discussed the classical methods for estimating θ in Chapter 5. Alternative Bayesian methods have been explored by Balding and Nichols (1997) and by Foreman et al. (1997).

Balding and Nichols modeled θ for the jth locus in the ith subpopulation by

$$\theta_{ij} \;\; = \;\; \frac{1}{1 + \alpha_i + \beta_j}$$

and assigned lognormal prior distributions to α and β. Allele proportions were assumed to have a Dirichlet distribution. The authors used data from a geographically-defined subpopulation and from a heterogeneous large population. Their estimate therefore referred to the relationship of alleles within each of the subpopulation and the large population when compared to alleles between the subpopulation and the large population.

Foreman et al. also allowed θ to vary over subpopulations and loci. They overcame the lack of subpopulation data by partitioning the population sample

We hope that we have done enough elsewhere in the book to show that there is no "right" number for this probability: the number that is given represents the scientist's reasoned and, in his opinion, balanced assessment of the weight of the evidence. Let us assume for the Gotham City example that the scientist calculates a match probability of one in a million, and therefore a likelihood ratio of one million.

RESULT OF A DATABASE SEARCH

In our example, we have explained that Mr. S. became a suspect for the rape of Ms. V. because of a lead from an informant. There is another kind of situation in which a suspect comes to notice: his profile is stored on a database of previous offenders. Should the fact that a suspect came to notice from a database search affect the evaluation of the weight of the scientific evidence? Although that is not the case in our example, this is a convenient point to discuss the question.

Let us imagine that the forensic science laboratory in Gotham City maintained a database of the profiles of previous offenders and that Mr. S. had come to notice because the profile from the vaginal swab was searched against the database and his profile was found to match. Let us assume that the database contained the profiles of N men and that Mr. S.'s was the only one to match. How would this affect the evaluation of the evidence? The NRC committee considered this issue and made this recommendation (Recommendation 5.1):

> When the suspect is found by a search of DNA databases, the random-match probability should be multiplied by N, the number of persons in the database.

So, if the Gotham City database contained 10,000 men, the recommendation would mean that the scientist would give a likelihood ratio of 100, rather than of one million. This would be a drastic dilution of the strength of the evidence and we need to look at the underlying logic rather more closely.

Recall, from Chapter 2, that in a case in which the evidence consisted of a stain left at the crime scene, the likelihood ratio took the form

$$\text{LR} = \frac{\Pr(E|H_p, I)}{\Pr(E|H_d, I)} \tag{9.1}$$

where $E = (G_S, G_C)$ and $G_C = G_S$. Balding and Donnelly (1996) make a useful distinction by referring to this as the *probable cause* LR; i.e., the LR where the suspect is arrested for reasons unconnected with his DNA profile. We have seen that, under certain assumptions, this LR reduces simply to the inverse of the match probability, $1/P$. Now consider that we have additional information: the suspect has been found as a result of a search of a database of N suspects. This particular suspect matched the crime genotype and the other $(N-1)$ did not. We have already seen how to deal with the event that the suspect and crime sample have the same genotype; let us use D to denote the additional information that the other members of the database did not. Then the LR in this case is

$$\text{LR} \;=\; \frac{\Pr(E, D|H_p, I)}{\Pr(E, D|H_d, I)} \tag{9.2}$$

Let us refer to this as the *database search* LR; we need to see how it differs from the probable cause LR. First, let us expand it using the multiplication law for probability as follows:

$$\text{LR} \;=\; \frac{\Pr(E|H_p, D, I)}{\Pr(E|H_d, D, I)} \frac{\Pr(D|H_p, I)}{\Pr(D|H_d, I)}$$

The first of these two ratios is very similar to the probable cause LR, Equation 9.1, save that the conditioning has been extended to include the information that none of the $(N-1)$ other suspects in the suspect's database has the crime profile G_C. The numerator is the probability of a match between suspect and crime stain given that the former left the latter: clearly the fact that there has been a database search does not affect this probability. The denominator is the match probability given the additional information that among $(N-1)$ other suspects there is no one with genotype G_C. It is clear that this extra information cannot increase the match probability; on the contrary, the fact that none of the other members of the database has that genotype increases confidence in its rarity and might tend to decrease the match probability. So this first ratio is approximately equal to (or slightly greater than) the probable cause LR of $1/P$. It follows that the database search LR is at least as large as the probable cause LR multiplied by the ratio R:

$$R \;=\; \frac{\Pr(D|H_p, I)}{\Pr(D|H_d, I)} \tag{9.3}$$

Now the problem is clearly specified: it is necessary to determine whether R is smaller than or greater than one. Let us first consider the numerator, which is the answer to the question

> If this suspect is the person who left the crime stain, what is the probability that none of the other $(N-1)$ suspects would match the crime stain?

Note that the conditioning does not include the crime stain genotype, so the question is concerned with the general discriminating power of the profiling system. We now write $\psi_{(N-1)}$ for the probability that none of $(N-1)$ innocent people

The first is easy to deal with because the scientist is not testifying to the validity of those circumstances: he is only repeating what he believes to be evidence to be presented by other witnesses. The second is actually an argument *in favor* of stating the perceived circumstances. It is essential that readers of the statement understand that the interpretation has taken place in a framework of circumstances. Furthermore, if the circumstances *do* change, then the scientist will need to review the interpretation and a sentence to this effect should, ideally, be included in the statement. It is a matter of judgment to decide which aspects of the circumstances need to be stated. Only those that are relevant to the interpretation should be included. In the present example these would be

- The alleged offense occurred in Gotham City.

- The offender was described as Caucasian and spoke with a local accent.

- The suspect has no close blood relatives in the city who would be considered suspects.

It is because of these features of the circumstances that the scientist has considered it most fitting to use a Caucasian database. The fact that his database was collected from Gotham City increases his confidence in its relevance because the offender appears to be local. It is because of the absence of blood relatives that he has calculated a match probability for an unknown person unrelated to the suspect, and his decision to work on the basis that the suspect and offender belong to the same subpopulation seems suitably cautious. The other aspects of the circumstances, such as the height, build, hair color, and age of the suspect and offender, while highly relevant to the deliberations of a court, do not appear relevant to the interpretation of the DNA evidence and can therefore be omitted from the statement.

Alternatives

We have seen that the first principle of interpretation states that it is not meaningful to address the uncertainty with regard to the truth of a proposition without considering at least one alternative proposition. In the present case there is one proposition that, at first sight at least, appears clearly defined. It is that of the investigator: Mr. S. raped Ms. V. It may be tempting for the scientist to take such an alternative as the first proposition for explaining the evidence, but a moment's reflection suggests that this might not be a wise course because the DNA profiling evidence cannot shed light on whether or not Ms. V. was actually *raped*: Mr. S. might later confirm that he and Ms. V. did have sexual intercourse but allege that it was with her full consent. In such an eventuality, the DNA evidence provides no assistance for weighing the prosecution and defense propositions against each other. Further thought suggests a proposition that is a stage or two removed from what prosecution will set out to prove. In this case it would seem sensible to suggest

H_p: The semen on the vaginal swab came from Mr. S.

The problem that we next face appears peculiar to the legal field because the alternative proposition should reflect the position of the defense but, of course, the defense in most jurisdictions are under no obligation to put forward any explanation for the evidence. The scientist must therefore anticipate a defense proposition and might decide to address something of the form: The semen on the vaginal swab did not come from Mr. S. Once again, reflection suggests that this may not be a helpful way of expressing the alternative. Let us recall what the scientist did. He calculated a match probability for an unknown person unrelated to Mr. S., and this must surely show that the alternative proposition being addressed is

H_d: The semen came from some unknown Caucasian man who was unrelated to Mr. S.

The scientist could, if he wished, make this still more specific to state that the unknown man came from the same subpopulation as Mr. S., but such a refinement is probably not needed and it may obscure clarity. The alternative propositions to be made clear in the statement are thus

H_p: The semen on the vaginal swab came from Mr. S.
H_d: The semen came from some unknown Caucasian man who was unrelated to Mr. S.

The scientist should also state his willingness to address other alternative propositions if necessary.

Evaluation

We have said that, given the defense proposition, the match probability is one in a million, and this is one way of expressing the strength of the evidence. The other way is to quote the LR of one million, because in this case the probability of the evidence given the prosecution proposition is one. In a case such as this, where the DNA evidence is very simple, there is no strong reason for choosing between the two methods. But this is not necessarily true when it comes to presenting the evidence in court, as we shall see shortly. Certainly, if the pattern of evidence is more complicated, such as when the crime sample is a mixture, then there is a 0.264 probability of there being more than one. The second flaw is more serious. Recall that we have calculated the match probability conditioned on the knowledge of the defendant's genotype. It is the probability that *a person other than the suspect* would have that genotype. If we want to talk about numbers of men we should be asking, given that we know that the defendant has that genotype, how many *other* men do we expect to find. Ignoring for the moment the issue of subpopulations and evolutionary relatedness, we return to the Poisson. The total number of other men in Gotham City is reduced by one, because we have set the defendant aside, but this obviously has no discernible effect on the calculation: the Poisson parameter is one, as before. Then the probabilities of 0, 1, 2, 3 . . . *other* men are as listed in the previous paragraph. So the probability that there is at least one other man in Gotham City is approximately $0.368 + 0.184 + 0.061 + \ldots = 1 - 0.368 = 0.632$.

The position with regard to the defendant is more easily seen if we use a Bayesian argument. If we consider it reasonable to consider all million males in the city as equally likely to have been the offender, in the absence of other evidence, the prior odds in favor of Mr. S. being the offender are a million to one against. The likelihood ratio for the DNA evidence is one million, so the posterior odds are one, or *evens*. Based on the uniform prior and the DNA evidence there is a 0.5 probability that Mr. S. was the assailant. The Bayesian arithmetic is just the same for the numbers cited in the G. Adams trial. Of course, even this is a weak method of reasoning. The idea that all males in the city should be regarded as equally likely suspects is quite unrealistic, as we discussed in Chapter 3. We return to such lines of argument later.

Concluding "It's Him."

It is almost inevitable, in any discussion of DNA evidence, that the question will be asked "Is it as good as fingerprints?" Courts throughout the world have been accustomed for decades to experts giving opinions of the form "This control print and this mark from the crime scene were made by the same person." It is not surprising that occasions have arisen where experts were prepared to give opinions of similar force with regard to DNA comparisons. For example, from *R. v. Deen*:

> Q: On the figures which you have established according to your research . . . what is your conclusion?
> A: My conclusion is that the semen has originated from Andrew Deen.

From the trial *R. v. Doheny*:

> Q: You deal habitually with these things, the jury have to say, on the
> evidence, whether they are satisfied beyond doubt that it is he. You
> have done the analysis, are you sure that it is he?
> A: Yes.

And from *R. v. G. Adams*:

> Q: Is it possible that the semen could have come from a different
> person from the person who provided the blood sample?
> A: It is possible but it is so unlikely as to really not be credible.

The ruling in *R. v. Doheny* quoted in the section on the transposed conditional now makes it clear that, in UK courts, scientists giving evidence on DNA are not to be asked to express an opinion as to whether a trace from a crime scene came from the defendant. DNA evidence alone, unlike all other kinds of transfer evidence, is peculiar in this regard. It might appear paradoxical that, in the single kind of transfer evidence for which meaningful statistical assessments can be made of the weight of evidence, the scientist is not to be permitted to express an opinion of origin. Vastly more is now known about DNA statistics than about handwriting, fingerprints, toolmarks, footwear, and so on, yet that evidence type alone is the one in which the scientist is not to express an opinion of origin. Later in the chapter we look at the reasons this should be so.

Size of Population Subgroups

We have seen that the Appeal Court in *R. v. Doheny* and *R. v. G. Adams* ruled that the expert was not to express an opinion on whether or not a given crime sample came from the defendant. The question then is: If the expert is not to give that guidance, how is the jury to reach a decision? The Appeal Court made some suggestions about how they may be assisted by the scientist, as follows:

> He will properly explain to the Jury the nature of the match ... be-
> tween the DNA in the crime stain and the DNA in the blood sample
> taken from the Defendant. He will properly, on the basis of empirical
> statistical data, give the Jury the ... frequency with which the match-
> ing DNA characteristics are likely to be found in the population at
> large. Provided he has the necessary data, and the statistical exper-
> tise, it may be appropriate for him then to say how many people
> with the matching characteristics are likely to be found in the United
> Kingdom–or perhaps in a more limited relevant subgroup, such as,
> for instance, the Caucasian sexually active males in the Manchester
> area. This will often be the limit of the evidence which he can prop-
> erly and usefully give. It will then be for the Jury to decide, having

evidence was to consider the non-DNA evidence numerically. The appropriate way to do this was by means of Bayes' theorem. The view of the statisticians who were advising prosecution was that this was too difficult an exercise for the jury, but if it were to be done, the prosecution offered its own experts to work with the defense expert in the preparation of a questionnaire for the jury to use. This was duly done. There were over 20 questions, directed at assessing the value of nonscientific aspects of the evidence. Here is an example, which is a pair of questions relating to the victim's description of her attacker:

> Bearing in mind that Ms. M. said that Mr. Adams appeared to be in his early forties what is the probability that she would say that her assailant was in his early twenties if Mr. Adams were indeed the assailant? What percentage of assailants in cases of this nature would be described as being in their early twenties?

The questions were designated by letters of the alphabet and questionnaire ended in a formula for combining the answers. The use of the questionnaire was explained in detail from the witness box by the defense statistician. Mr. Adams was again convicted and, once more, appealed. The following are extracts from the judgment of the court that considered the second appeal (*R. v. D. Adams*, 16 October 1997, CA 96/6927/Z5):

> If ... the jury concluded that they did accept the DNA evidence wholly or in part called by the Crown, then they would have to ask themselves whether they were satisfied that only X white European men in the United Kingdom would have a DNA profile matching that of the rapist who left the crime stain. It would be a matter for the jury, having heard the evidence, to give a value to X. They would then have to ask themselves whether they were satisfied that the defendant in question was one of those men. They would then go on to ask themselves whether they were satisfied that the defendant was the man who left the crime stain, bearing in mind on the facts of this case the obvious discrepancies between the victim's description of her assailant and the appearance of the appellant, ... Of course, it is a matter for the jury how they set about their task, and it is no part of this court's function to prescribe the course which their deliberations should take. But consideration of this case along the lines indicated would in our judgment reflect a normal course for a properly instructed jury to adopt. It is the sort of task which juries perform every day ... as they are sworn to do.

> We are very clearly of opinion that in cases such as this, lacking special features absent here, expert evidence should not be admitted

to induce juries to attach mathematical values to probabilities arising from non-scientific evidence adduced at the trial.

The appeal was dismissed.

It might be thought that the rulings were against the Bayesian view of scientific evidence. This is not correct: the judges were ruling against its application for the evaluation of *nonscientific* evidence. This, in fact, corresponds to the advice given to the prosecution by its own experts in both trials. But the judgment is, ultimately, unsatisfactory. The jury faced a difficult task in this case–powerful evidence supporting Mr. Adams' involvement, yet persuasive evidence from the victim herself supporting his innocence. The judgment clearly states that the jury should not be expected to tackle this problem logically. However, they are expected to assign a value X to the number of men in the country who would match the crime profile, though we have already seen that this line of reason can be problematic for a scientist, let alone a lay person. If the jury decided that $X = 1/10$, for example, how is this to be related to the conflict between victim's description of her attacker and the appearance of the defendant? This is hardly "the sort of task which juries perform every day." The root problem is that the existing legal system exerts powerful forces against carrying through the most appropriate procedure effectively. Notwithstanding the efforts put into the design of the questionnaire and its explanation to the jury, it must still have remained a puzzling exercise to them. However, if sufficient time and resources had been forthcoming, then a computer-based dialogue could have been produced. If free dialogue between the expert and jury had been possible, then the system could have been tailored to the intellectual capabilities of each individual jury member. Although such a solution is well within the bounds of today's technology, the procedural difficulties would be immense because of the nature of the established institution.

INDIVIDUALIZATION AND IDENTIFICATION

It will be useful at this point to look at some general issues relating to the process of identification and then to consider the process of fingerprint comparison.

Individualization

In the forensic sphere, the words *identity* and *identical* tend to be misused. Examiners sometimes give opinions of the kind "these two marks are identical." This is not correct because any entity can only be identical with itself. Two marks, whether they are from the same finger, from the same item of footwear or made by the same tool, cannot be *identical* and indeed they will inevitably be *different* in detail. Two different entities cannot be identical to each other because they are each unique. This applies not only to so-called "identical twins" but also to all of the grains of sand on a beach. Note also that a DNA profile is a manifestation of a complex biological/physical/chemical process and two DNA profiles cannot be identical to each other, even if they have come from the same person. The fact that we choose to summarize each profile by a set of numbers and that two profiles have the same sets of numbers merely means that they are indistinguishable from each other using the measuring system that we have chosen.

The issue for the forensic scientist is not "Is this profile unique?" (it is) or "Are these two things identical?"(they can't be) but "Is there sufficient evidence to demonstrate that they originate from the identical source?" We notice that it is widespread practice in the forensic field to refer to the process that leads to the answer "yes" to this question as *identification*. Kirk (1963) pointed out that the word *individualization* was more appropriate in this regard; indeed, he defined criminalistics as the "science of individualization." Nevertheless, we must bow to what has become general usage–certainly in the fingerprints field–and refer to a categorical opinion of identity of source as an "identification."

DNA versus Fingerprints

Earlier we raised the question "Is a DNA profile as good as a fingerprint?" It is important that we should understand a fundamental difference between the processes of inference that are pursued in the two fields, which was concisely explained by Stoney (1991):

> Fingerprint comparisons have the colloquial specificity of absolute identification, but a completely different [compared to DNA profiling] philosophy for achieving it. Although the study of fingerprint variation is founded on scientific observations, the process of comparison and the conclusion of absolute identity is explicitly a subjective process. The conclusions are accepted and supported as subjective; very convincing, undoubtedly valid, but subjective. In fingerprint comparisons, the examiner notes the details in the patterns of ridges. Beginning with a reference point in one pattern, a corresponding point in a second pattern is sought. From this initial point the examiner then seeks neighboring details that correspond in their form, position and orientation. These features have an extreme variability that is readily appreciated intuitively, and which becomes objectively obvious upon detailed study. When more and more corresponding features are found between two patterns, scientist and lay person alike become subjectively certain that the patterns could not possibly

be duplicated by chance.

What has happened here is somewhat analogous to a leap of faith. It is a jump, an extrapolation, based on the observation of highly variable traits among a few characteristics, and then considering the case of many characteristics. Duplication is inconceivable to the rational mind and we conclude that there is absolute identity. This leap, or extrapolation, occurs (in fingerprinting) without any statistical foundation, even for the initial process where the first few ridge details are compared. A contrast with our DNA individualization process is important because we in no way approach DNA evidential interpretation in the same way we approach fingerprints. We hold fingerprint specificity and individuality up as our ideal, yet this ideal is achieved (and can only be achieved) through a subjective process that we patently reject when applied to DNA. With DNA typing, as in conventional serological typing, we view our increasing evidential value as a step-wise process. We detect a series of traits, each one of which is, to some degree, rare. This leads to the inference of smaller and smaller joint probabilities and a conclusion that the combined type would be very very rare.

Stoney contrasts the undisputed subjectivity of a fingerprint comparison with what he sees to be the objectivity of a DNA statistic. Yet we have seen that this objectivity is itself an illusion because it exists only within a framework of assumptions. In the individual case it is for the scientist to judge the validity of those assumptions and to carry out whatever calculations he considers necessary given the case circumstances. In the wake of the Doheny/Adams appeal ruling in the UK, there has been a tendency for courts to seek a "statistical probability" or a "mathematical probability" in the mistaken belief that such numbers exist independently of human judgment.

We should be in no doubt about the degree of certainty implicit in a fingerprint identification. The expert is, in effect, saying "I am certain that this latent mark and this control print were made by the same person and *no amount of contrary evidence will shake my certainty.*" Or, to look at this from a Bayesian perspective, no matter how small the prior odds are, the likelihood ratio is so large that the posterior odds approach infinity. Stoney sees that a fingerprint identification is based on a "leap of faith," and he is quite correct to conclude that such a leap of faith has nothing to do with scientific principles. It is that leap of faith that characterizes the essence of a conclusion of identity of source and, as he points out, that is a fundamental difference between fingerprint evidence and DNA evidence. Stoney's "leap of faith" is equivalent to attaining an infinite likelihood ratio: this kind of belief cannot derive from any *scientific* process.

Ultimately, it must always be the jurors, or other triers of fact, whose belief in the proposition of an identical source that matters. The question is about the role that the scientist plays in determining that state of belief in the juror's mind. With conventional evidence types it has long been accepted in the courts that it is right and proper for the scientist to give his view on the proposition of an identical source and then it is a matter for the juror to decide on his confidence in the expert's judgment. However, the judgment in *R. v. Doheny* and *R. v. G. Adams* means that, for British scientists at least, that must not be done with DNA evidence.

So, in considering the question "Is it as good as a fingerprint?" we must recognize that a fingerprint identification is based on a process that is quite different in nature from that which we follow in interpreting a DNA match. The fingerprint identification means that the expert has reached a characteristic mental state of complete certainty based on the skilled and complex comparison that he has made. No juror is competent to attempt that comparison. As Stoney says, the expert does not *prove* individuality, he becomes mentally convinced by it. The issue is only proved when the court decides that he is competent to give that opinion and the jury decides that he can be believed.

With DNA, on the other hand, once the genotypes of the crime profile and the suspect have been determined, the comparison is trivial–any juror can see whether or not they are the same. Whereas the fingerprint expert does not consciously dissociate the two components, numerator and denominator, of the likelihood ratio, with a DNA match we are generally happy with the notion that the numerator is one and the assessment of the weight of the evidence comes down to considering the magnitude of the denominator. Once we assign a number to the denominator then we must recognize that we have given the court something that they may choose to work with without our assistance. Certainly, the idea that the scientist has some particular power to take that number and take a step equivalent to the Stoney "leap of faith" is misconceived. If he really wishes to emulate the fingerprint expert, he must say "that match probability is so small that no amount of contrary evidence will shake me from the opinion that the crime sample was left by the defendant."

The expression "DNA fingerprinting" fosters an unrealistic impression of the technology, and it should not be encouraged in forensic circles.

Independence Across Loci

There is a key statement in the 1996 NRC report:

We foresee a time when each person can be identified uniquely (except for identical twins).

The report contains language to the effect that, when DNA profiles match at a large number of loci, it is not reasonable to believe that they come from different people. This is based on our understanding that each person is genetically unique, identical twins excepted. The NRC statement reflects the widespread view that individualization through DNA profiling is a matter of testing at a sufficient number of loci. This is an understandable position to take: it appears to be inarguable that the more matching loci, the better the evidence. But how do we combine likelihood ratios from the different loci? Clearly, we would prefer to multiply them and justify this by an independence assumption. Providing the likelihood ratios are moderate enough that matches can be found in a database, then we can investigate the robustness of the assumption by suitable experiments based on between-person comparisons as we described in Chapter 5. As a rough guide, it seems reasonable that a likelihood ratio of, say, one million can be presented credibly if the scientist can quote a between-person experiment based on at least a million comparisons. The experiments of Lambert et al. (1995) and Evett et al. (1996) describe millions of comparisons, the former based on four-locus RFLP data and the latter on four-locus STR data. Larger experiments on RFLP data were conducted by Risch and Devlin (1992).

As we test more and more loci, we find larger and larger likelihood ratios for matching profiles, and we face two problems. First the credibility problem: "How can you quote such large numbers based on such relatively small databases?" Second, the closely related problem of testing robustness because we are combining the evidence by a method whose robustness we cannot possibly test. Certainly there are strong *a priori* reasons to believe that if all the loci are well separated throughout the genome, then the weight of evidence increases as more and more loci are added. This belief is strengthened by the decreasing proportion of matching profiles found in between-person experiments as more loci are considered. However, the computed statistic, as Stoney (1991) pointed out, is a personal statement of belief, and most certainly not an objective "statistical probability." As an illustration, imagine that we have a 12-locus match for which we have computed a likelihood ratio of 10 billion. We now test an extra four loci, all of which match. Is it now meaningful to say that the likelihood ratio is 100 trillion? Whether or not it is meaningful to quote such an extravagant number, we must be in no doubt that its magnitude depends on independence assumptions to a measure that we cannot possibly support by data. So when we add more loci, the notion that the evidence is becoming more and more compelling is intimately related to personal belief. There is nothing wrong with this, and indeed there is nothing new about it because it is the notion that fingerprint experts invoke as they find more and more points of comparison. The same applies to handwriting compari-

Table A.2 Chi-square values that are exceeded with specified probabilities.

d.f.	0.995	0.990	0.975	0.950	0.900	0.100	0.050	0.025	0.010	0.005
1	0.00	0.00	0.00	0.00	0.02	2.71	3.84	5.02	6.63	7.88
2	0.01	0.02	0.05	0.10	0.21	4.61	5.99	7.38	9.21	10.6
3	0.07	0.12	0.22	0.35	0.58	6.25	7.81	9.35	11.3	12.8
4	0.21	0.30	0.48	0.71	1.06	7.78	9.49	11.1	13.3	14.9
5	0.41	0.55	0.83	1.15	1.61	9.24	11.1	12.8	15.1	16.7
6	0.68	0.87	1.24	1.64	2.20	10.6	12.6	14.4	16.8	18.5
7	0.99	1.24	1.69	2.17	2.83	12.0	14.1	16.0	18.5	20.3
8	1.34	1.65	2.18	2.73	3.49	13.4	15.5	17.5	20.1	22.0
9	1.73	2.09	2.70	3.33	4.17	14.7	16.9	19.0	21.7	23.6
10	2.16	2.56	3.25	3.94	4.87	16.0	18.3	20.5	23.2	25.2
11	2.60	3.05	3.82	4.57	5.58	17.3	19.7	21.9	24.7	26.8
12	3.07	3.57	4.40	5.23	6.30	18.5	21.0	23.3	26.2	28.3
13	3.57	4.11	5.01	5.89	7.04	19.8	22.4	24.7	27.7	29.8
14	4.07	4.66	5.63	6.57	7.79	21.1	23.7	26.1	29.1	31.3
15	4.60	5.23	6.26	7.26	8.55	22.3	25.0	27.5	30.6	32.8
16	5.14	5.81	6.91	7.96	9.31	23.5	26.3	28.8	32.0	34.3
17	5.70	6.41	7.56	8.67	10.1	24.8	27.6	30.2	33.4	35.7
18	6.26	7.01	8.23	9.39	10.9	26.0	28.9	31.5	34.8	37.2
19	6.84	7.63	8.91	10.1	11.7	27.2	30.1	32.9	36.2	38.6
20	7.43	8.26	9.59	10.9	12.4	28.4	31.4	34.2	37.6	40.0
21	8.03	8.90	10.3	11.6	13.2	29.6	32.7	35.5	38.9	41.4
22	8.64	9.54	11.0	12.3	14.0	30.8	33.9	36.8	40.3	42.8
23	9.26	10.2	11.7	13.1	14.8	32.0	35.2	38.1	41.6	44.2
24	9.89	10.9	12.4	13.8	15.7	33.2	36.4	39.4	43.0	45.6
25	10.5	11.5	13.1	14.6	16.5	34.4	37.7	40.6	44.3	46.9
26	11.2	12.2	13.8	15.4	17.3	35.6	38.9	41.9	45.6	48.3
27	11.8	12.9	14.6	16.2	18.1	36.7	40.1	43.2	47.0	49.6
28	12.5	13.6	15.3	16.9	18.9	37.9	41.3	44.5	48.3	51.0
29	13.1	14.3	16.0	17.7	19.8	39.1	42.6	45.7	49.6	52.3
30	13.8	15.0	16.8	18.5	20.6	40.3	43.8	47.0	50.9	53.7
40	20.7	22.2	24.4	26.5	29.1	51.8	55.8	59.3	63.7	66.8
50	28.0	29.7	32.4	34.8	37.7	63.2	67.5	71.4	76.2	79.5
60	35.5	37.5	40.5	43.2	46.5	74.4	79.1	83.3	88.4	92.0
70	43.3	45.4	48.8	51.7	55.3	85.5	90.5	95.0	100.4	104.2
80	51.2	53.5	57.2	60.4	64.3	96.6	101.9	106.6	112.3	116.3
100	67.3	70.1	74.2	77.9	82.4	118.5	124.3	129.6	135.8	140.2

Table A.3   Two thousand random digits.

	5	10	15	20	25	30	35	40	45	50
1	30246	86149	45548	80480	85924	02411	46456	23952	55145	18300
2	02806	20733	30853	08034	21238	39933	90958	87912	82486	96960
3	84868	17425	91536	08208	44761	40101	74109	08696	73249	10885
4	65043	86343	36953	04658	42008	84984	49584	53872	52737	24217
5	59792	12608	73246	57277	29384	02608	78779	59311	08421	72618
6	29008	02705	38780	09675	32573	74039	85654	12731	36846	21341
7	74800	20695	99211	38699	28454	21400	11524	81212	55327	93367
8	45715	29459	60745	64762	81553	00401	21852	65586	51269	73813
9	70056	78054	16563	32244	81117	26808	94318	00873	00154	81690
10	30072	38515	52181	21872	17193	57361	16000	51633	70345	48725
11	19490	00789	48629	84877	18858	73868	05461	57469	58009	23998
12	79558	05067	71799	72777	45475	39847	14211	09764	38988	94242
13	18072	34286	46778	95843	31600	57151	89995	58712	46820	81464
14	09933	43223	27657	00697	84736	96171	18120	74205	86558	72670
15	68396	26040	44227	73036	11903	59352	73105	88131	25523	48473
16	76023	01624	74545	18347	66573	79479	24729	98822	93629	72477
17	52257	64895	96218	45817	93951	30547	93632	21510	17326	95743
18	27531	76301	89645	24680	93157	56419	92677	05539	81408	37221
19	17406	68465	66526	13785	92655	25101	95658	54255	07336	17904
20	87810	83955	12467	83985	39484	80179	96878	67468	16173	29937
21	01109	37024	09219	04303	65058	07201	50126	56572	97194	99595
22	67362	79269	61078	70412	89414	45697	17368	48025	41999	45286
23	38002	58000	50220	34603	73647	06894	84712	52922	73303	22802
24	60044	14258	82451	24551	14223	77858	61729	69565	62211	90630
25	55818	55177	80015	88181	96369	57150	37206	02369	18457	29621
26	82646	47169	71375	65259	13194	59086	81076	08421	47402	25764
27	47133	75669	28424	83710	21907	46183	21782	04475	88099	33155
28	62065	06444	34797	56543	90176	41665	53588	71810	26557	83977
29	52765	89407	17693	33927	97348	72061	14231	12340	44493	64194
30	68651	84960	60535	51369	08459	97693	31991	37836	37247	50762
31	74437	48122	89309	16025	06062	10840	22809	28746	30682	48082
32	49051	14405	76357	57632	46511	00666	09647	61493	66875	29164
33	95023	70370	60841	58975	63641	71478	48327	82378	17689	49232
34	19358	28765	57897	93980	61832	10202	79416	40162	85205	87337
35	95489	73778	86660	39424	89005	68527	85534	77132	95116	65790
36	07758	15002	18281	35417	07440	56681	31392	91160	85337	79306
37	27602	69590	13299	50384	25829	85184	89773	97149	16399	41287
38	75864	68804	37205	39021	67019	38964	62848	40359	22254	54700
39	47313	78390	64495	14918	97584	73636	55745	33592	16050	86578
40	13406	80860	65073	73149	74121	97974	60190	50744	52846	91673

Appendix B

SOLUTIONS TO EXERCISES

CHAPTER 1

Exercise 1.1

$$
\begin{aligned}
\Pr(H_1|E) &= \frac{1}{3} \\
\Pr(H_1 \text{ or } H_2|E) &= \frac{1}{3} + \frac{1}{3} = \frac{2}{3} \\
\Pr(\bar{H}_1|E) &= 1 - \frac{1}{3} = \frac{2}{3} \\
\Pr(H_1|0) &= 0
\end{aligned}
$$

Exercise 1.2

Let C denote Caucasian, H denote highland, L denote Celtic language, and E denote the information that the person is selected at random from the voter registration list. Then

$$
\begin{aligned}
\Pr(C, H, L|E) &= \Pr(L|C, H, E) \Pr(H|C, E) \Pr(C|E) \\
&= 0.75 \times 0.2 \times 0.8 \\
&= 0.12
\end{aligned}
$$

Exercise 1.3

Let G denote the event that a person has the required genotype, Ca the event that a person is Caucasian, Mo the event that a person is Maori, and Pa the event that a person is Pacific Islander. Then

$$
\begin{aligned}
\Pr(G) &= \Pr(G|Ca) \Pr(Ca) + \Pr(G|Mo) \Pr(Mo) + \Pr(G|Pa) \Pr(Pa) \\
&= 0.013 \times 0.8347 + 0.045 \times 0.1219 + 0.039 \times 0.0434 \\
&= 0.018
\end{aligned}
$$

Exercise 1.4

a.

$$\Pr(\text{Both dice are even}) \quad = \quad \frac{1}{2} \times \frac{1}{2} = \frac{1}{4}$$

$$O(\text{Both dice are even}) \quad = \quad \frac{\frac{1}{4}}{1 - \frac{1}{4}} = \frac{1}{3}$$

i.e., 3 to 1 against.

b.

$$\Pr(\text{Both dice show a six}) \quad = \quad \frac{1}{6} \times \frac{1}{6} = \frac{1}{36}$$

$$O(\text{Both dice show a six}) \quad = \quad \frac{\frac{1}{36}}{1 - \frac{1}{36}} = \frac{1}{35}$$

i.e., 35 to 1 against.

Exercise 1.5

a.

$$\Pr(H) \quad = \quad \frac{19}{1 + 19} = \frac{19}{20} = 0.950$$

b.

$$\Pr(H) \quad = \quad \frac{0.2}{1 + 0.2} = \frac{1}{6} = 0.167$$

c.

$$\Pr(H) \quad = \quad \frac{1000}{1 + 1000} = \frac{1000}{1001} = 0.999$$

d.

$$\Pr(H) \quad = \quad \frac{1/1000}{1 + 1/1000} = \frac{1}{1001} = 0.001$$

Exercise 1.6

a.

$$\Pr(D|I) \quad = \quad 1/10,000 = 0.0001$$

b.

$$\Pr(E|D, I) \quad = \quad 0.99$$

Exercise 6.3

The numerator for LR is

$$\Pr(G_C = A_iA_j|G_M = A_iA_j, G_{AF} = A_iA_j, H_p) \;=\; \frac{1}{2}$$

Because there is doubt as to which of the child's alleles is maternal and which is paternal, we need to sum over both possibilities, as stated in the text. The denominator for LR is

$$
\begin{aligned}
\text{Den.} \;=\; & \Pr(A_M = A_i|G_M)\Pr(A_P = A_j|G_M, G_{AF}, H_d) \\
& + \Pr(A_M = A_j|G_M)\Pr(A_P = A_i|G_M, G_{AF}, H_d) \\
=\; & \frac{1}{2}[\Pr(A_P = A_j|G_M, G_{AF}, H_d) + \Pr(A_P = A_i|G_M, G_{AF}, H_d)]
\end{aligned}
$$

As in the text, we need to consider the possible genotypes for MM, and all we know is that this person must have at least one of alleles A_i and A_j. The analog of Table 6.5 is:

G_{MM}	$\Pr(G_{MM})$	T_1	T_2	T_3 $A_P = A_i$	$A_P = A_j$
A_iA_i	p_i^2	$\frac{1}{2}$	$\frac{p_i^2}{p_i+p_j}$	$\frac{3}{4}$	$\frac{1}{4}$
A_iA_j	$2p_ip_j$	$\frac{1}{2}$	$\frac{2p_ip_j}{p_i+p_j}$	$\frac{1}{2}$	$\frac{1}{2}$
A_jA_j	p_j^2	$\frac{1}{2}$	$\frac{p_j^2}{p_i+p_j}$	$\frac{1}{4}$	$\frac{3}{4}$
A_iA_k $k \neq i,j$	$2p_ip_k$	$\frac{1}{4}$	$\frac{p_ip_k}{p_i+p_j}$	$\frac{1}{2}$	$\frac{1}{4}$
A_jA_k $k \neq i,j$	$2p_jp_k$	$\frac{1}{4}$	$\frac{p_jp_k}{p_i+p_j}$	$\frac{1}{4}$	$\frac{1}{2}$

$$\Pr(G_M = A_iA_j|G_{AF} = A_iA_j) = (p_i + p_j)/2$$
$$T_1 = \Pr(G_M = A_iA_j|G_{MM}, G_{AF} = A_iA_j, H_d)$$
$$T_2 = \Pr(G_{MM}|G_M, G_{AF})$$
$$T_3 = \Pr(A_P|G_{MM}, G_{AF} = A_iA_j, H_d)$$

Averaging over the two values for A_M, the denominator becomes

$$
\begin{aligned}
\text{Den.} \;=\; & \frac{p_i^2}{2(p_i + p_j)} + \frac{p_ip_j}{p_i + p_j} + \frac{p_j^2}{2(p_i + p_j)} \\
& + \sum_{k \neq i,j} \frac{3p_ip_k}{8(p_i + p_j)} \\
& + \sum_{k \neq i,j} \frac{3p_jp_k}{8(p_i + p_j)}
\end{aligned}
$$

$$= \frac{3 + p_i + p_j}{8}$$

and LR is $4/(3 + p_i + p_j)$.

Exercise 6.4

Under the proposition H_p that the sample is from the missing person, the probability of the evidence is:

$$
\begin{aligned}
\Pr(E|H_p) &= \Pr(G_P, \{G_S\}, G_M, G_C, G_X | H_p) \\
&= \Pr(G_C|G_X, G_M)\Pr(G_X, G_M, \{G_S\}|G_P)\Pr(G_P) \\
&= \Pr(G_C|G_X, G_M)\Pr(G_M)\Pr(G_X, \{G_S\}|G_P)\Pr(G_P) \\
&= \Pr(G_C|G_X, G_M)\Pr(G_M)\Pr(G_P)\sum_{G_F}\Pr(G_X, \{G_S\}|G_P, G_F)\Pr(G_F)
\end{aligned}
$$

where

$$
\begin{aligned}
\Pr(G_C = A_3A_5|G_X = A_3A_3, G_M = A_5A_6) &= 1/4 \\
\Pr(G_M) &= 2p_5p_6 \\
\Pr(G_P) &= 2p_3p_4 \\
\Pr(G_X, \{G_S\}|G_P = A_3A_4, G_F = A_2A_3) &= 1/1024 \\
\Pr(G_X, \{G_S\}|G_P = A_3A_4, G_F = A_2A_4) &= 0 \\
\Pr(G_F = A_2A_3) &= 2p_2p_3 \\
\Pr(G_F = A_2A_4) &= 2p_2p_4
\end{aligned}
$$

Therefore

$$\Pr(E|H_p) = p_2p_3^2p_4p_5p_6/512$$

Under the proposition H_d that the sample is not from the missing person, the probability of the evidence is:

$$
\begin{aligned}
\Pr(E|H_p) &= \Pr(G_P, \{G_S\}, G_M, G_C, G_X|H_p) \\
&= \Pr(G_X)\Pr(G_M)\Pr(G_C|G_M)\Pr(\{G_S\}|G_P)\Pr(G_P) \\
&= \Pr(G_X)\Pr(G_M)\Pr(G_P)\sum_{G_F}\Pr(G_C|G_M, G_F)\Pr(\{G_S\}|G_P, G_F)\Pr(G_F)
\end{aligned}
$$

where

$$
\begin{aligned}
\Pr(G_X) &= p_3^2 \\
\Pr(G_M) &= 2p_5p_6 \\
\Pr(G_P) &= 2p_3p_4 \\
\Pr(G_C = A_3A_5|G_M = A_5A_6, G_F = A_2A_3) &= 1/4
\end{aligned}
$$

natural selection in random mating populations. Evolution 13:561–564.

Li, C.C. 1988. Pseudo-random mating populations. In celebration of the 80th anniversary of the Hardy-Weinberg law. Genetics 119:731–737.

Li, Y.J. 1996. *Characterizing the Structure of Genetic Populations*. Ph.D Thesis. N.C. State University.

Lindley, D.V. 1977. A problem in forensic science. Biometrika 64:207–213.

Lindley, D.V. 1982. Coherence. Pp. 29–31 in Kotz, S., N.L. Johnson and C.B. Read (Eds.) *Encyclopedia of Statistical Sciences*, Vol. 2. Wiley, New York.

Lindley, D.V. 1991. Probability. Pp. 27–50 in Aitken, C.G.G. and D.A. Stoney (Eds.) *The Use of Statistics in Forensic Science*. Ellis Horwood, New York.

Maiste, P.J. and B.S. Weir. 1995. A comparison of tests for independence in the FBI RFLP data bases. Genetica 96:125–138.

Morris, J.W., R.A Garber, J. d'Autremont and C.H. Brenner. 1988. The avuncular index and the incest index. Pp. 607–611 in *Advances if Forensic Haemogenetics 1*. Springer-Verlag, Berlin.

Mosteller, F. and D.L. Wallace. 1964. *Applied Bayesian and Classical Inference - the Case of the Federalist Papers*. Springer-Verlag, New York.

National Research Council. 1996. *The Evaluation of Forensic DNA Evidence*. National Academy Press, Washington, DC.

Nichols, R.A. and D.J. Balding. 1991. Effects of population structure on DNA fingerprint analysis in forensic science. Heredity 66:297–302.

O'Hagan, A. 1994. *Kendall's Advanced Theory of Statistics 2B*, Wiley, New York.

Risch, N. and B. Devlin. 1992. On the probability of matching DNA fingerprints. Science 255:717–720.

Robertson, B. and G.A. Vignaux. 1995. *Interpreting Evidence: Evaluating Forensic Science in the Courtroom*. Wiley, Chichester.

Roychoudhury, A.K. and M. Nei. 1988. *Human Polymorphic Genes*. World distribution. Oxford University Press, Oxford.

Stoney, D.A. 1991. What made us ever think we could individualize using statistics? J. For. Sci. Soc. 31: 197–199.

Thompson, W.C. and E.L. Schumann. 1987. Interpretation of statistical evidence in criminal trials - The prosecutors fallacy and the defense attorneys fallacy. Law and Human Behavior 11; 167–187

Tribe, L.H. 1971. Trial by Mathematics: Precision and ritual in the legal process. Harvard Law Review 84:1329–1393.

Vogel, F. and A.G. Motulski. 1986. *Human Genetics*, Second Edition. Springer-Verlag, New York.

Walker, R.H., R.J. Duquesnoy, E.R. Jennings, H.D. Krause, C.L. Lee, and H. Polesky (Eds.). 1983. *Inclusion Probabilities in Parentage Testing*. American Association of Blood Banks. Arlington, VA.

Walsh, K.A.J, J.S. Buckleton and C.M. Triggs. 1994. Assessing prior probabilities considering geography. J. Forensic Sci. Soc. 34: 47-51.

Wambaugh, J. 1989. *The Blooding*. William Morrow, New York.

Weir, B.S. 1994. Effects of inbreeding on forensic calculations. Ann. Rev. Genet. 28:597–621.

Weir, B.S. 1996. *Genetic Data Analysis II*. Sinauer, Sunderland, MA.

Weir, B.S. and C.C. Cockerham. 1984. Estimating F-statistics for the analysis of population structure. Evolution 38:1358–1370.

Weir, B.S. and I.W. Evett. 1992. Whose DNA? Am. J. Hum. Genet. 50:869.

Weir, B.S. and I.W. Evett. 1993. Reply to Lewontin. Am. J. Hum. Genet. 52:206.

Weir, B.S., C.M. Triggs, L. Starling, L.I. Stowell, K.A.J. Walsh and J.S. Buckleton. 1997. Interpreting DNA mixtures. J. Forensic Sci. 42:113–122.

Wright, S. 1951. The genetical structure of populations. Ann. Eugen. 15:32–354.

Wright, S. 1965. The interpretation of population structure by F-statistics with special regard to systems of mating. Evolution 19:395-420.

Zaykin, D., L. Zhivotovsky and B.S. Weir. 1995. Exact tests for association between alleles at arbitrary numbers of loci. Genetica 96:169–178.

Author Index

Subject Index